AF659136

# ETHICS OF ECOLOGY AND ENVIRONMENT

Ethics of

# ECOLOGY AND ENVIRONMENT

*S. Subbarao*

**RAJAT PUBLICATIONS**
**NEW DELHI - 110 002 (INDIA)**

**RAJAT PUBLICATIONS**
4740/23, Ansari Road, Daryaganj
New Delhi- 110 002 ( India)
Phones : +91-11-23267924, +91-9811484973
Email: rajat_publication@yahoo.com

*Ethics of Ecology and Environment*

© *Reserved*

Edition 2026

ISBN 978-81-7880-017-2

[The responsibility for facts stated opinion expressed or conclusions reached and plagiarism, if any, in this volume is entirely that of the Editor. The publisher bears no responsibility for them whatsoever.]

PRINTED IN INDIA

---

Published by Mrs. Seema Wasan for **Rajat Publications**

# PREFACE

The newly emerging concept of ecology and environment has moved into the mainstream of public life as a major national and international concern. This provides important opportunities for environmental sociologists to contribute to the understanding and solution of these problems. An interesting insight concerns here, how each of the several identities of the Arctic, (e.g., as homeland for the natives, as a "colony" exploited for its natural resources, and as the last wilderness) results in a distinctive approach to the human—environment relationships.

In the area of human-environment relationships, there is need for studies on conflict resolution to strike a balance between commercial needs and the interests of subsistence hunters, sportsmen, and conservation. The second area pertains to community viability, for which a systematic approach is needed to help in development of physical and social services infrastructure to meet the special climatic needs of the region. A final area pertains to the study of the impact of rapid social change (single-industry cash economy, the snowmobile revolution), which is exacting a heavy price from the local inhabitants in terms of higher rates of alcoholism, suicide, stress, lonelines, accidents, and violence.

The commitment to protect growth or a certain way of life has led conventional human ecologists and the neo-Malthusians to disregard the minimum well-being or sheer survival of a rest of humankind. In fact, Hawley admits that while competition, resulting from demand exceeding the carrying capacity, may accounts for the exclusion of some contestants from access to their share of a limited resource. It does not shed any light on what happens to the exclude members of a population after their exclusion. This serious neglect of the concern for the under dog and the undermined is matched by the self-admitted tendency of the human ecologists not to confront policy matters directly.

Besides introductory outline, this work has been classified in two parts--first part deals with ethics of ecology in following chapters: Historical concepts of ecological balance; Moral dilemma of ecological crisis; Ecological consciousness from religious to technological thought; Transcending ecology: Christian and Marxist approach; and Eco-philosophy of Buddhism; however, second Part throws light on ethics of environment in following chapters: Nature and scope of environmental ethics; Ethics of life protection; Western view of environmental ethics; Hindu ethics of environment; environmental economy and Justice; and Assessment of environmental impact.

In completing this work, I have taken support from various sources including books and journals. I am thankful to all those great scholars whose works have been cited or substantially made use of in this work.

**—Editor**

# CONTENTS

# 1

# INTRODUCTION

The most distinguishing feature of the twentieth century is that a single species, humanity, has become the creator of its own future. Through, the mastery of science and its applications, humans possess the ability to change the environment of Planet Earth, and can do so within the lifetime of a single species member.

From its genesis as a chunk of molten rock and gas some 4.5 billion years ago, the earth's climate or topology has undergone successive upheavals and extinctions. Ice and fire have sculpted the shape of the planet from age to age as mountains have risen from the deep and continents have sunk to the ocean floor. This struggle of life and death is spectacularly illustrated by the demise of the great dinosaurs during the Cretaceous period, some 136 to 65 million years ago. There is much speculation as to the specific cause of their death, though there is little doubt it was brought about by some sudden shift in the environment. The most plausible theory is that a mammoth meteorite crashed on the earth, sending up billows of dust which blocked out the sun, thus destroying the plants on which the dinosaurs survived.

The shocking fact is that similar waves of extinction are breaking out all over the globe at this very time, but in the present instance the destroyer is not some visitor from outer space or some balancing act of nature, but humans themselves. Twentieth century *Homo sapiens* has become a force of nature, and like the sun, the moon, the wind, and water, people exercise a power over nature which can reduce it to the fate of the dinosaur.

The environmental crisis in the United States arises out of the dilemmas of prosperity. In his study of America in the mid-nineteenth century, Alexis de Tocqueville observed: "Their ancestors gave them the love of equality and of freedom; but God Himself gave them the means of remaining equal and free by placing them upon a boundless continent." Indeed, the natural abundance of this country has made it a "promised land" of biblical proportions—a cornucopia of riches which have, no doubt, helped shape the national character of its people. In the words of a recent president, "The breadth and variety and beauty of our land, the richness of our mines and soil and forests and water, the favourable nature of our climate—all of these natural factors have provided a setting in which the optimism, the ingenuity and drive of the American people thrive and grow and are rewarded."

The price of plenty has also meant that it has been impossible for Americans to imagine a time when its "spacious skies" could be polluted, and its "alabaster cities" be reduced to asphalt jungles. Since, these problems were not anticipated, neither were we prepared to prevent them. "Even as we are heirs to the products of our forefathers' genius, so our generation has also inherited the results of past carelessness."

Actually, the problem is one of more than "carelessness." It is a defect which arises out of the strengths of American society—its science, democracy, individuality, and prosperity. Broadly, it is a flaw of our civilisation; and that ultimately boils done to how we, as westerners, perceive ourselves in relation to nature.

This chapter addresses the challenge coming out of America's abuse of the environment: its failure to respect the delicate web we share with other forms of life, and its neglect of humane design in manufactured artifacts, most conspicuously in the area of urban planning. Our purpose is not to indulge in doom and gloom, but to go beyond mere survival to finding new ways of bringing quality into our lives and restoring the balance of nature.

We begin by isolating the main agent of environmental destruction. The picture is much more complicated, but if we are to narrow the field to a single cause it would have to be technology, or better stated, our handling of technology.

Modern technology is the special child of Western culture; that is to say, it required a particular ethos and ideology in which to take birth and flourish. Western philosophy set the stage with a built-in dualism between spirit and matter reaching back to Plato and earlier to the Orphic mysteries. This strain of classical thinking was passed on to the modern West through René Descartes (1596-1650), French philosopher and scientist. In his scientific investigations, he followed the methods laid down by Francis Bacon. His philosophy was almost completely dualistic. His view of the physical world was that of a machine, wholly different from the mind. Only through the intervention of God were the world and mind brought together. As the "father of modern philosophy," Descartes had an immense impact on the development of European thought, especially through the Cartesian philosophers, who chiefly studied the problem of the relation of body and soul, of mind and matter.

During the Enlightenment, the mechanistic interpretation of the world was symbolised by the workings of a watch. Watches came in diverse forms and shapes, but their movements were all predetermined. Once God, conceived as the heavenly Watchmaker, had created it, the machine was able to run as determined. The sharp metaphysical line between humans and their environment was continued in the thinking of Berkely, Kant, Hegel, and Sartre. None of these philosophers conceived of the subhuman world in terms that would entitle it to fundamental human concern.

Western religion has followed a similar path. The three major Western religions—Christianity, Judaism, and Islam—teach that the universe is the creation of God, who has given human beings a special place in that creation. The book of Genesis (1:28-30) describes the position of persons in relation to the rest of creation: And God said to them, "Be fruitful and multiply, and fill the earth and subdue it; and have dominion over the fish of the sea and over the birds of the air and over every living thing that moves upon the earth." And God said, "Behold, I have given you every plant yielding seed which is upon the face of all the earth, and every tree with seed in its fruit: you shall have them for food. And to every beast of the earth, and to every bird of the air, and to everything that creeps on the earth, everything that has the breath of life, I have given every green plant for food."

The story of Creation is followed by the story of the Fall. Later, a wholesale purge of wickedness became necessary, and God sent the Flood, which exterminated not only the majority of the human population but innocent subhuman species as well. God then made a convenient with his servant Noah in which man's earlier dominance was reaffirmed, and he was also permitted the liberty of a carnivorous diet. Genesis 9:1-3 states: And God blessed Noah and his sons, and said to them, "Be fruitful and multiply, and fill the earth. The fear of you and the dread of you shall be upon every beast of the earth, and upon every bird of the air, upon everything that creeps on the ground and all the fish of the sea; into your hand they are delivered. Every moving thing that lives shall be food for you; and as I gave you the green plants, I give you everything.

The pattern of Western philosophy and religion to distinguish sharply between humans and nature, and thereby to objectify nature so that it is seen from the outside, was extended by Western science. The separation became the attitudinal basis for every-increasing knowledge of nature by science. In Western thought, nature was descralised, and therefore considered an object fit to be studied by separating part from part, limb from limb. "This superlatively effective way of discovering solidly verifiable truths tends, precisely because it is sharply focused, to ignore whatever lies outside its periphery of attention."

This phenomenon goes a long way toward explaining why Western culture has dominated the field of science for four hundred years. This has worked both as an asset for Western culture and also as a liability. "Because our strength is derived from the fragmented mode of our knowledge and our action, we are relatively helpless when we try to deal intelligently with such unities as a city, an estuary's ecology, or 'the quality of life.'"

To the extent that the goal of science is knowledge, it must be freed of any direct blame for the environmental crisis. On the other hand technology, which is oriented to action, derives its power from science and applies it in its projects. According to Max Ways, "Although, its categories are not the same as those of science, technology in its own way is also highly specialised, directed toward narrowly defined aims. As its power rises, technology's 'side effects,' the consequences lying outside its tunneled field of purpose, proliferate with disastrous consequences to the environment—among other unintended victims."

To conclude this train of thought, our argument is that technology is the central agent behind the destruction of the environment, and that the West's innovations in this field derive from attitudes which have been shaped by the dominant religion-philosophical currents of Western culture. The principle of separation or differentiation has produced much scientific discovery and advance, but its very virtue—progress through isolation, individuation, and fragmentation—has also been its weakness in that its tunnel vision has proven myopic in viewing the environment as a whole.

There is one danger in tracing the roots of the environmental crisis to its cultural roots in religion. In his celebrated essay, "The Historic Roots of our-Ecologic Crisis," Lynn White asserts that "what people do about ecology depends on what they think about themselves in relation to things around them. Human ecology is deeply conditioned by beliefs about our nature and our destiny—that is, by religion. The danger in this sort of claim is that it makes religion the primary conditioner of human interaction with the environment.

The model which White proposes has three progressive steps: (1) Judeo-Christian tradition, (2) science and technology, and (3) environmental degradation. The picture is more complicated. Certainly, religion has a basic role to play, since it accounts for the way we perceive ourselves, our fellows, and the world around us. But Western civilisation has also been conditioned by the institutions of democratisation and capitalism, aided by science and technology. These political and economic forces have produced wealth, individual ownership, urbanisation, and population increase. Each one of these sources has taken more from the environment than it has put in, and thereby has been responsible for degrading it. The principle of separation, the method used by science to separate one suspect of nature from another for the purpose of study, has equally been at work in the operations of our political and economic institutions, and this has only provided us with short-term, piecemeal approaches to our environmental problems. We now turn to the major problems which make up the environmental crisis.

## ECOLOGICAL PROBLEMS

In a rare departure from its tradition of naming a "Man of the Year," *Time* magazine designated Endangered Earth as Planet of the year for 1988. Like God warning Noah of the coming flood, this was the year that Planet Earth spoke with tremors which were heard around the globe.

In the United States drought stalked the land from California to the Deep South, depleting the grain harvest by 31 per cent, and killing thousands of cattle. The combined, effects of heat waves and lack of rain ignited forest fires, with the largest conflagration in Yellowstone National Park. Beaches were fouled with garbage, raw sewage, and medical wastes, showing the limited capacity of the ocean to absorb pollution.

Similar despoliation of the oceans occurred in areas of the Mediterranean, the North Sea, and the English Channel, making beaches unfit for recreation. Killer hurricanes roared through the islands of the Caribbean and sent thousands in Bangladesh to a watery grave. In Soviet Armenia, 55,000 persons perished in a gigantic earthquake. There were grave warnings of the further depletion of the ozone layer, and all the while tropical forests, which house half the earth's plant and animal species, were being destroyed at the speed of one football field per second. Most of these evils had been going on for a long time, and some of the worst disasters apparently had nothing to do with human behaviour. Yet, this year's

(1988) bout of freakish weather and environmental horror stories seemed to act as a powerful catalyst for worldwide public opinion. Every one suddenly sensed that this gyrating globe, this precious repository of all the life that we know of, was in danger. No single individual, no event, no movement captured imaginations or dominated headlines more than the clump of rock and soil and water and air that is our common home.

With each passing year, threats to the survival of Planet Earth have increased with nothing less than mind-boggling rates. Following are the chief problem areas; biodiversity, global warming, pollution, waste, overpopulation, and the nuclear epidemic.

## The Problem of Biodiversity

Mother Nature never puts all of her eggs in one basket: diversity is the key to survival. Variety is not just the spice of life, but its sum and substance. "Complex beyond understanding and valuable beyond measure, biological diversity is the total variety of life on earth."

During the time when humans were hunters ad gatherers, their sustenance depended on biodiversity. This dependence changed when society looked for its livelihood, first to agriculture, and then to industry. Now, with the research knowledge that biological diversity is crucial for the environment which sustains life, people are being roused to ensure its conservation.

Our knowledge of the total count of life with which we share this planet is still shrouded in mystery, but what we do know is that biodiversity is vanishing at an in calculable speed. Eugene Linden tells a story of what is happening in Brazil, but which is actually a global phenomenon. Before Brazil's great land rush, the emerald rain forests of Rondonia state were an unspoiled showcase for the diversity of life. In this lush territory south of the Amazon, there was hardly a break in the canopy of 200-ft.-tall trees, and virtually every acre was alive with the cacophony of all kinds of insects, rids and monkeys. Then, beginning in the 1970s, came the swarms of settlers, slashing and burning huge swaths through the forest to create roads, towns and fields. They came to enjoy a promised of land, but they have merely produced a network of devastation. The soil that supported a rich rain forest, is not well suited to corn and other crops, and most of the newcomers can eke out only an impoverished, disease-ridden existence. In the process, they are destroying an ecosystem and the millions of species of plants and animals that live in it. An estimated 20 per cent of Rondonia's forest is gone, and at present rates of destruction it will be totally wiped out within 25 years.

Like Brazil, India is also a land of rich diversity. Its geographic location makes it "one of the world's top twelve 'megadiversity nations.'" But the subcontinent's biodiversity is being dangerously eroded. One cause for India's loss is earlier British policy, which "drastically reoriented the resource-use pattern, largely exporting teak, tea, and indigo outside the locality. The prices were greatly depressed. After Independence, this system of undervaluing biological resources was taken to extremes. Thus bamboo, of great utility to the rural people, has been made over to the paper industry at royalties of Rs. 1.50 a tonne, when the price for basket weavers was over Rs. 3000 a tonne."

A second cause for the erosion of biodiversity in India is the scant attention paid to the quality of life among its rural and tribal peoples. "In pre-British India, vast tracts were available to fulfil the biomass needs of the local people. Parts of such lands were set aside as 'sacred groves' for total protection of biological communities, while others were treated as community woodlots from which regulated harvests were carried out." A few of these systems survive today in remote areas such as Mizoram, but they were deliberately and effectively destroyed by the British, who granted no rights but only some privileges to the local communities over common lands. "As a consequence such lands became 'no man's lands' abused by all but protected by none. This situation has worsened after Independence. All this has affected the local people such as the tribals." The stories of India's and Brazil's battles for biodiversity are sadly repeated all over the globe.

Biodiversity is analysed in three ways: by ecosystems, species, and genetic diversity. Ecosystems are the communities in which organisms live an move and have their being. They comprise forests, wetlands, mangroves, and reefs. Of these, tropical forests are of highest value. They cover only 7 per cent of the earth's surface, but they are the habitat for 50 to 80 per cent of its species. Tropical forests are now only half the original size. In 1990, deforestation reduced the at aggregate further by an area the size of Washington State.

Of all nations, Brazil has the most extensive range of tropical forests, housing the highest number of species. By 1987, as a result of government action, economic shifts, and climatic changes, the massive deforestation of earlier years was brought under control. "Moreover, with nearly 90 per cent of its groves still standing, by national or international standards the Brazilian Amazon is relatively untouched. Brazil's most endangered ecosystems are its unique coastal forests. Logging and agriculture and urban expansion have destroyed more than 95 per cent of the once-vast costal rain forests and the coniferous Araucaria forests of southern Brazil." Given the fact that only 5 per cent of the world's tropical forests are protected by their governments, "the state is set for mass extinctions."

Temperate rain forests are in even greater danger than the tropical forests. Half of the original 13 million hectares have fallen prey to logging. In the United States, only pockets of the old-growth rain forests survive in the Pacific Northwest.

In India, trees are among the most important of natural resources, but this has not lessened the pace of deforestation cross the land. In Punjab, forest cover declined from120,000 hectares to 49,000 hectares between 1972 and 1982. In Rajasthan, some three hundred mines extracting marble, limestone, soapstone, silica, bauxite, and granite have suddenly brought an army of eight thousand labourers into the forests of Alwar district, with predictable damage. Madhya Pradesh is undergoing rapid denudation of its forest cover "from the notional 35 per cent of the area to 19 per cent." In Assam, deforestation is "the greatest threat to the environment in the region as a whole," producing leser rainfall and severe floods caused by hillside erosion. In Tripura, the Gumti dam project has ousted some 2,600 tribals from their homes, forcing them to seek livilihood in the upper catchment areas of the Gumti and to resort to the practice of *jhum* (slash-and-burn) agriculture, causing tremendous damage to the trees.

Uttar Pradesh is high among the regions hardest hit by deforestation. "From time immemorial, the Himalayas have played a crucial role in maintaining the ecological balance, particularly in the North. In literature and legend, these pristine heights have served for millennia as the playground of the Hindu gods, and their exquisite beauty has inspired generations of holy men to seek heaven on earth, but like the cedar groves of Lebanon, soon the only record of the Himalayan forests may be found in sacred texts.

Wetlands, like forests, are productive habitats for biodiversity. They are home to countless varieties of fish and waterfowl, and have natural mechanisms which control water flows and cleanse the system of sediments and pollution.

Among India's most picturesque lakes, Renuka, situated in Himachal Pradesh, is a crown jewel. But Lake Renuka is dying. In 1989, the Ministry of Environment and Forests designated it the state's only "natural wetland" and accorded it national status. But the state's rhetoric has fallen short of conservation efforts to save the lake from being choked by elephant grass and heavy salutation during monsoons.

The greatest loss of wetlands, exceeding 90 per cent, has occurred in New Zealand, Australia, and in California. The blights of heavy industry, urbanisation, agricultural use, over-grazing, and logging are among its enemies. Canada has a quarter of the world's wetlands, which fare the best.

## HUMAN ECOLOGY AND ENVIRONMENT

The term *ecology* comes from the Greek word *oikos* (house) and, significantly, has the same Greek root as the word *economics,* from *oikonomos* (household manager). Ernst Haeckel, the German biologist who coined the word *ecology* in 1868, viewed ecology as a body of knowledge concerning the economy of nature, highlighting its roots in economics and evolutionary theory. He defined ecology as the study of all those complex interrelations referred to by Darwin as the conditions of the struggle for existence.

Ecologists like to look at the environment as an ecosystem of interlocking relationships and exchanges that constitute the web of life. Populations of organisms occupying the same environment (habitat) are said to constitute a community. Together, the communities and their abiotic environments constitute an ecosystem. The various ecosystems taken together constitute the ecosphere, the largest ecological unit. Living organisms exist in the much narrower ranger of the biosphere, which is said to extend a few hundred feet above the land or under the sea. On its fragile film of fair, water, and soil, all life is said to depend. For the sociologist, the most important ecological concepts are diversity and dominance, competition and cooperation, succession and adaptation, evolution and expansion, and carrying capacity and the balance of nature. Over the years, three distinct approaches have come to characterise the field of human ecology.

The Chicago sociologists Louis Wirth, Robert Ezra Park, Ernest W. Burgess, and Roderick McKenzie are recognised as the founders of the human ecological approach in sociology. In the early decades of the twentieth century, American cities were passing through a period of great turbulence due to the effects of rapid industrialisation and urbanisation. The urban world, with its fierce competition for territory and survival, appeared to mirror the very life-world studied by the plant ecologists. In their search for the principles of order, the human ecologists started looking at the process of city development in terms of the process of "succession," which involved an orderly sequence of invasions and displacements leading to a climax or equilibrium state. In their hands, human ecology became synonymous with the ecology of space. Park and Burgess identified the "natural areas" of land use, which come into existence without a preconceived design. Quite influential and popular for a while was the "Burgess Hypothesis" regarding the spatial order of the city as a series of concentric zones emanating from the central business district. However, Hawley (1984) has pointed out that with urban characteristics now diffused throughout society, one in effect deals with a system of cities in which the urban hierarchy is cast in terms of functional rather than spatial relations.

Park (1936) identified the problematics of human ecology as the investigation of the processes by which biotic balance and social equilibrium are maintained by the interaction of the three factors constituting what he termed the "social complex" (population, technological culture [artifact], and nonmaterial culture [custom and belief] with the fourth, natural resources of the habitat. Essentially, the same factors reappear as the four POET variables (population, organisation, environment, and technology) in Otis Dudley Duncan's Ecological Complex, indicating its point of contact with the early human ecology (Duncan 1964). In any case, it was McKenzie who, by shifting attention from spatial relations to the analysis of sustenance relations, provided the thread of continuity between the two approaches. His student Amos Hawley, who has been the "exemplar" of conventional human ecology since the 1940s, defines it as the attempt to deal holistically with the phenomenon of organisation.

Hawley (1986) views the ecosystem as the adaptive mechanism that emerges out of the interaction of population, organisation, and the environment. Organisation is the adaptive form that enables a population to act as a unit. The process of system adaptation involves members in relations of interdependence to secure sustenance from the environment. Growth is the development of the system's inner potential to the maximum size and comlexity afforded by the existing technology for transportation and communication. Evolution is the creation of higher potential for resumption of system development through the incorporation of new information that enhances the capacity for the movement of people, materials, and messages. In this manner, the system moves from simple to more complex forms.

Through the use of the ecological concept of "expansion," Hawley (1979) has applied his framework to account for the growth phases that intervene between stages of development. The evolution of a system takes place when its scale and complexity do not go hand in hand. Thus, an imbalance between population and the carrying capacity of the environment may create external pressures for branching off into colonies and establishing niches in a new environment. However, when complexity and scale do advance together, the normal condition for growth or expansion arises from the colonisation process itself. Hawley identifies the technology of movement as the most critical variable. It determines accessibiity and, therefore, governs the spread of settlements, changes in hierarchy and division of labour, and the creation of interaction networks among the settlements. According to Hawley, the process can work on any scale and is limited only by the level of technological development of communication and transportation.

Hawley (1979) points out that while expansionism in the past relied on political domination, its modern variant aims at structural covergence along economic and cultural axes to obviate the need for direct rule by the centre. In this way, free trade and "resocialisation" of cultures and whole populations into standardised organisational forms and procedures created a far more efficient and cost-efective global reach. At the same time, as polities try to bring larger portions of system territories under their jurisdictions, the management of scale becomes highly problematical. Lacking a single supranational polity, the international pecking order is then subject to increasing instability, challenge, and change.

The system ends to return to scale and the costs of administration mount. This results in decentralisation and some local autonomy. But improvements in technology of movement start the growth process all over again. The picture thus is that of a global system thoroughly interlinked by transportation and communication networks. Hawley believes that the growth of social systems has now reached a point at which the evolutionary model has lost its usefulness in explaining cumulative change. However, since a single world order has only a small tolerance for errors, this situation harbors the grave danger that a fatal error may destroy the whole system.

The real irony of this relentless global expansion elaborated by Hawley lies, however, in the coexistence of extreme opulence and affluence of the few with stark poverty and misery of the majority at home and abroad. The large metropolitan centres provide a very poor quality of life. The very scale of urban decay underscores the huge problems facing the city—congestion, polluted air, untreated sewage, high crime rates, dilapidated housing, domestic violence, and broken lives. One therefore needs to ask: What prospect does this scale and level of complexity hold for the future?

Industrial and industrialising nations are now beset with more or less the same devastating problems of air, land, and water pollution and environmental destruction. Large numbers of lakes and rivers that were not naturally eutrophic have now become so as a result of pollution and chemical run-offs. In the United States, Love Canal and Times Beach, Missouri, made headlines in the 1980s as much as Chernobyl did in 1986 in the Soviet Union. Sulphur dioxide emissions from industrial and power plants cause acid rain that inflicts irreparable damage on buildings, monuments, marine life, trees, and plants. Over sixty thousand synthetic chemicals are now on the market, of which a sizable number contaminate the environment and pose health hazards. Over half a million tons of toxic wastes are produced each year in the United

States, where the five-year cost of cleaning the nuclear waste, which remains dangerously radioactive for thousands of years, may well exceed thirty billion dollars. The soil and the lake water and groundwater near nuclear power and weapons plants are heavily contaminated with such toxins as mercury, arsenic, and many types of solvents, and with deadly radioactive materials such as plutonium, tritium, and strontium-90. The contamination is so bad in eight states that huge tracts of land are said to be totally unfit for human habitation and pose serious health hazards for the surrounding communities.

The environmental destruction is far more serious and widespread in Eastern Europe and the Soviet Union. These countries are the site of some of the world's worst pollution. Lakes and rivers are dead or dying. Water is so contaminated in some areas that it is undrinkable. Chemical run-off and sewage and wastewater dumping have created serious groundwater contamination. Lignite (brown coal), the major source of energy for industry and homes in some of these nations, is responsible for heavy concentration of sulphur dioxide and dust in the air that has caused serious respiratory problems and additional health damage. The haze-covered cities are an environmental disaster. According to Worldwatch estimates, the Soviet Union alone accounts for a fifth each of global carbon dioxide and sulphur dioxide emissions—the former are implicated in global warming; the latter are the principal ingredient of acid rain. As a result, environment and politics are becoming an explosive mix and the source of action and instability in the region. Zhores Medvedev believes the Soviet Union is losing its forests at the same rate as rain forests are being destroyed in Brazil. Between 1960 and 1989, it lost more land through land and environmental degradation than the combined total cultivable area in Belgium and

Ireland. Besides, the effects of Chernobyl and other major nuclear disasters, it lost more agricultural land and pasture land to radioactive contamination than the total cultivated acreage in Switzerland. Experts fear that from twenty-five thousand to seventy-five thousand persons in Russia and Europe may die prematurely from cancer and other reflects of the radiation fallout from Chernobyl, history's worst nuclear disaster. A quarter to a third of the forests in eastern Europe show signs of dying from air pollution. These environmental problems are thus not only transboundary, they also cut across ideological labels. Between them, the United States and the Soviet Union produce more than 40 per cent of the carbon dioxide emissions every year.

In non-western nations, a million people suffer acute poisoning and twenty thousand persons die every year from pesticides. Pesticides are a major source of environmental and health problems in the United States as well. But the United States alone exports over half a billion pounds of pesticides that are restricted or banned for domestic use. The ecology, natural environment, and resources of these non-Western nations are being destroyed and contaminated at a frightening rate. Irreversible damage is being done by large-scale destruction of rain forests and the intensive use of marginal lands and by the imbalances that results from population pressures and the practices of multinational firms and national elites. Desertification now threatens a third of the earth's land surface. Poverty, hunger, starvation, famine, and death are endemic throughout much of the world.

## ENVIRONMENTAL SOCIOLOGY AND THE HUMAN ECOLOGY

The mounting public concern during the 1970s about fuel shortages, oil spills, nuclear powerplant accidents, acid rain, dying lakes, urban smog, famine and death in the Sahel, rain forest destruction, and the like made social scientists realise that overexploitation of the ecosystem may destroy the very basis of our planetary survival. Many environmentalists blamed the dominant social paradigm of industrial societies for the destruction of the fragile balance among the components of the ecological complex.

The antiecological worldview of the dominant social paradigm, which was shaped by the experience of a seemingly limitless frontier and of extraordinary abundance, is said to be a major reason for the neglect of the physical environment by American sociologists. Another reason is related to the Durkheimian emphasis on explaining social facts by other social facts. Together with the general aversion of sociologists to any form of determinism, these reasons ruled out the due consideration of environmental factors even by human ecologists. William Catton and Riley Dunlap (1978) point out that the exaggerated emphasis by conventional human ecologists on culture, science, and technology as "exceptional" human achievements has led to the illusion that humans are "exempt" from bioecological constraints to which all species are subject. This awareness has led Catton and Dunlap to develop the fields of new human ecology and "environmental sociology" to deal with the reciprocal interaction between human activities and the physical environment. They believe that the POET model, broadened to include the role of human agency and culture, provides a useful analytical framework for grounding environmental sociology in the ecological perspective. In addition, they have identified the important subfield of the sociology of environmental issues, which is concerned with the study of the environmental movement, wild land recreation, and resource management problems and the like, but mainly within the sociological tradition.

In a comprehensive review of the new field, Buttel (1987) has made two essential points. He argues that both the cleavage between the conventional and the new human ecology, and the conflict-versus-consensus cleavage of mainline sociology, continue to be important. Buttel would also like to shift the focus of environmental sociology from the imbalance of population and resources, emphasised by Catton, to the reality of the unequal distribution of these resources. Allan Schnaiberg's idea of the "treadmill of production" (1980), which emerges from a dialectical relationship between economic growth and ecological structures, points to the need for focusing on production institutions as the primary determinants of economic expansion and for incorporating a conflict dimension in environmental analysis. Buttel's own work in environmental sociology draws upon the "political economy tradition" of the neo-Marxists and the neo-Weberians. Catton's major contributions, on the other hand, are in the neo-Malthusian tradition.

Thus, while the problem of order created by the harsh realities of industrial life and expansionism defined the central problematics of sociology and conventional human ecology, the problem of survival now defines the central problematics of environmental sociology and the new human ecology: to the earlier question of how social order is possible is now added the more urgent concern with survival itself.

A "new urban sociology" conflict paradigm in the political economy tradition has been put forward by M. Gottdiener and Joe Feagin (1988) as an alternative to the conventional and the new human ecology paradigms. Societies, in their view, are neither mere population aggregates nor unified biotic communities, but are specified by their mode of production. Crisis tendencies and profit generation constitute the core of societal development, which is dominated by the capital accumulation process. Thus, to take one example, conventional human ecologists like to regard central-city restructuring as a consequence of adaptation to increasing population size and the growing complexity of social organisation. They then relate these changes to the size of the metropolitan hinterland. The new urban paradigm,on the other hand, emphasises the impact of a global economy, the multinational corporations, the shift to functional specialisation in world-system financial and administrative activities, the constant subsidisation by the state, the efforts of pro-growth coalitions, and changes in labour force requirements leading to some renovation and central city gentrification. The following are some of the basic questions that the new urban sociology paradigm seeks to answer: What is the character of power and inequality? How do they relate to "ecological"patterns? How do production and reproduction processses of capital accumulation, as well as the processes of crisis adjustment, manifest themselves in sociospatial organisation?

## RELATION BETWEEN POPULATION AND ENVIRONMENT

For ecology, the most salient aspect of the population—environment relationship is the way it affects human survival and the quality of human life. Under the impact of the interlocking crisis of overpopulation, resource depletion, and environmental degradation, issues of sustainability and survival have come to occupy centre stage. Corresponding to the main approaches in human ecology, three broad positions may be identified for discussion: the pro-growth (expansionist), the neo-Malthusian, and the political economy perspectives. The discussion of these positions is followed by a consideration of the Brundtland Report, issued by the World Commission on Environment and Development, and of the traditional-

Gandhian view of the ecological crisis. Extended treatment of the issues involved may be found in Catton (1980), Schnaiberg (1980), Humphrey and Buttel (1982), Redclift (1987), Mellos (1988), and de la Court (1990).

*The Pro-Growth (Expansionist) Perspective.* In expansionist thinking, scale, complexity, and acceleration—that is, the constant broadening of the limits of the maximum permitted by prevailing circumstances—mark the human-environment encounter. The basic conviction is that man through his unaided reason can comprehend and control the processes of the world. This is the meaning of Hawley's apparently guarded conviction that evolution as a mode of change has disappeared from human social systems that is, that by grasping the very main-spring of evolution, man has already seized "the tiller of the world," in the words of Teilhard de Chardin. In responding to the concern of the neo-Malthusians about population outrunning resources, Hawley, points to the inherently expansive nature of populations, technology, and organisation, which through resource substitutions, intensive use and exploitation of land and resources, and the exploration and discover of new frontiers and new resources, can allay concerns regarding environmental finitude. In his view, the problem of food and resource shortages is really rooted in poverty rather than in resource scarcity. In short, Hawley believes that industrial systems have no known upper limits on either the number of specialisations or the size of the populations that can be supported. Similar pro-growth sentiments are expressed by other expansionist thinkers.

While Colin Clark directly links population numbers to power, Herman Kahn (1974) views population increase as a necessary stimulus to economic growth and believes the earth can easily support fifteen billion people at twenty thousand dollars per capita for a millennium. In fact, he believes that the wider the gap between the rich and the poor, the more the riches will percolate downward. In any case, he is unconvinced that the rich would agree to part with their income to ensure a more equitable distribution of wealth. Roger Revelle (1974) believes the earth can actually support nearly thirty times the present population in terms of food supplies, and that it would take almost 150 years to hit that mark. While economic development is necessary to provide people with the basis to control their fertility, Revelle is certain the world would drown in its own filth if most of the people in the world were to live at Western standards. Finally, the post industrial sociologist Daniel Bell (1977) is convinced that economic growth is necessary to reduce the gap between the rich and the poor nations. He has little doubt that the "super-productivity society," with less than 4 per cent of its labour force devoted to agriculture, could feed the whole palatine of the United States, and most of the world as well. In his opinion, pollution exists because the market principle has never been applied to the use of collective goods. Actually, bell suggests that the government itself could utilise the market to demand a public accounting from all parties on issues of public interest, levy effluent charges for pollutants, and bring effective compliance through the price mechanism.

However, while corporations have shown greater sensitivity and self-regulation, there is evidence that the attempts to enforce the "polluter pays" principle are likely to be resisted or the costs passed on to the public. The negative impact of governmental policies that alleviate energy and resource scarcities is more likely to be felt at the lower socio-economic levels (Morrison 1978). Dunlap (1979) presents evidence to show that the effects of pollution and the costs of cleaning the environment are borne disproportionately by the poor and may actually serve to reinforce class inequalities.

*The Neo-Malthusian Perspective.* The neo-Malthusian perspective is based on the well-known *Malthusian law* of a geometric increase in population far outstripping the arithmetic increase in food supplies in the absence of constraints. The disparity between the two growth rates poses a perpetual threat to the human prospect and serves as a natural brake on unlimited population increase by triggering war, widespread famine, disease, and death. In place of these harsh "positive checks." Malthus (1766-1834) wished to substitute the "preventive check" of "moral restraint," not the artificial methods of birth control and abortion. The crucial linking concept between population and environment is the biological concept of "carrying capacity." The carrying capacity for humans is the maximum population that can be supported indefinitely by a particular environment under exploitation by specified technology and organisation. With each technological break through, humans have so far been able to raise their carrying capacity through extending their territorial and environmental reach, which now goes to the ends of the globe. A more or less stable population of five hundred million in 1650 jumped to a billion in 1850. In 1930, it was two billion; and in 1975, four billion. Now humans number 5.2 billion, with prospects of the earth's population doubling about once every thirty years. Three-fourths of the earth's inhabitants live in non-Western nations. The 1990 U.S. population of 250 million is expected to jump to 300 million in 30 years.

Compounding the environmental effects of the poverty-stricken and "food hungry" populations of the world are the impacts of massive consumption and pollution of the "energy hungry" nations. The latter rise sharply with even a slight growth in population of Western nations, where one-quarter of the world's population is responsible for over 85 per cent of worldwide consumption of natural resources and the environmental sinks. Within the United States a bare 6 per cent of the world's population consumes over half of the world's nonrenewable resources and over a third of all the raw materials produced. Miller believes that the real threat to our life-support system, therefore, comes not from the poor but from the effluent megaconsumers and megapolluters who occupy more space, consume more of natural resources, disturb the ecology more, and directly and indirectly pollute the land, air, and water with ever-increasing amounts of thermal, chemical, and radioactive wastes. While the Club of Rome and the other neo-Malthusians give a grace period of thirty or so years, Catton believes we have already overshot the maximum carrying capacity, and are now on a catastrophic downward crash course. In any case, he is convinced that our best bet would be to act *as if* a crash were imminent and to take advance measures to minimize its impact.

However, these premises of Malthusian theory have not stood the test of time. The social and economic forces unleashed by the Industrial Revolution not only telescoped the doubling of human population within a shorter time span, they also brought about ever-rising material standards of living due to astronomical increases in the scale and seed of agricultural and industrial production in the advanced nations. Worldwide, the annual rate of population growth fell to 1.8 per cent in 1990, and is expected to stabilise around 1.5 percent by the year 2000. Annual food production in the developing countries is increasing by 4.4 per cent, over twice the rate of pollution growth. What has caused the huge increase in numbers in poor nations is the fact that while the birth rate has declined, the death rate has gone down as well. In Western nations, on the other hand, fertility rates registered a dramatic decline between the two world wars, boomed thereafter to everybody's disbelief, and have again hit such low levels that population replacement has become a major concern for many industrial nations. The Western nations are thus said to have completed a "demographic transition." In the initial stage, population is kept stable as high birthrates are balanced by high death rates due to the operation of the Malthusian positive checks. In the second stage, population increases rapidly as death rates decline but birth rates soar due to an increase in food supply and improved health care. In the third stage, birth rates decline more rapidly than death rates due to the use of family planning and birth control measures. Finally, low birth rates and low death rates lead to a stabilisation of the population at a lower level. A decline in fertility rates to a level below replacement (1.3 per woman) in industralised countries such as Germany and Denmark is said to constitute the current "second demographic transition." France and the Soviet Union have officially encouraged childbearing to counter falling birth rates. Thus, in the present era, the challenges and opportunities opened up by declining fertility rates and record food surpluses have totally eclipsed the Malthusian specter for the rich nations while again casting its shadow on the poor. In any case, population is but one factor, albeit a most significant one, in the complex interaction between social structure and the environment.

To revert to the neo-Malthusian argument: the tragedy of numbers is compounded by the "free rider," who derives personal benefits from the collective efforts of others, and the more serious "tragedy of the commons", where each herdsman will add cattle without limit, ignoring the costs imposed on the others and must, therefore, degrade the land held in common. Hardin believes that the administrative system, supported by a common standard of judgement and having access to "mutual coercion mutually agreed upon," can avoid the collective tragedy. However, unless the users jointly own the commons and exercise joint control over the resources and their use, Hardin's cure may place them at the mercy of the custodians, leaving them worse of individually and collectively. The threat of imminent ethnocide facing the remaining native population is a case in point.

On the other hand, the "development" of poor nations has created a new set of claimants for the resources needed to maintain the high material standard of living of affluent nations. As the poor nations begin to assert control over their own resources, try to set terms of their exchange, or resist outside pressures to transform them into "environmental preserves" or the "global commons," the prospects of conflict, particularly over critical mineral and energy resources, are greatly magnified. Amartya Sen (1981) has looked at the famine situation as essentially a "crisis of entitlement," not so much because there is lack of food but because the many are denied any claims to it because of their lowly position in society. In the West, the entitlement revolution has entailed huge welfare expenditures, which could be financed either by economic growth or by direct redistribution of income. For Bell and the neo-Malthusians, the latter is out of the question. There is thus a natural progression from Hardin's "commons" view to his "lifeboat ethics." While he admits it is possible to significantly increase food supply to match the demand, he is convinced that every Indian life saved would only diminish the quality of life of those aboard the "lifeboat" and of their descendants. The situation, he feels, demands a world authority to take control of reproduction and resource use. In a piece published in 1969 in the *Stanford Alumni Almanac,* and appropriately titled" The Immorality of Being softhearted," Hardin is quite clear that food would be the worst thing to send to the poor. Nothing short of the final solution will do. "Atomic bombs would be kinder. For a few moments, the misery would be acute, but it would soon come to an end for most of the people, leaving a very few survivors to suffer thereafter." These solutions, which bring about decimation of entire populations, have been called *ecofascist.* Such sentiments are by no means uncommon among the neo-Malthusians.

The "tragedy of the commons" is really the tragedy of individualism from which all constraints of private and common morality have been removed. However, others have been quick to point to the equal or far greater extent of environment pollution and ecological destruction in socialist countries as one more evidence of the inevitable convergence of capitalism and socialism! Many environmental problems are clearly transideological and transnational. Acid rain, oil spills, destruction of the ozone layer, threat of global warming—all a call for common responsibility and joint regulation. Ironically, it appears that the expressed concern about the destruction of the global commons though overpopulation or industrial pollution is seldom matched by a parallel commitment by powerful nations to preserve or clean up the environment or provide support for international population control efforts. Instead, one witnesses a mad scramble to divide up the remaining oceanic and other planetary resources without regard to equity, ecology, or environment. As a result, the hair and the oceans, as well as the forests and lands of other nations, are being over exploited or used as garbage and toxic dumps with impunity.

Of no small consequence globally is the environmental impact of waste, widespread corruption at all levels, hoarding and price-fixing, and poor storage, distribution, and transportation networks. B.B. Vohra, a top Indian government official, has pointed out how "formidable mafias based on a triangular alliance between the corrupt bureaucrat, the corrupt politician and the corrupt businessman emerged in all [Indian] States and became a most powerful threat to the conservation of the country's tree cover". When one adds to this list the role of political and economic elites and multinational corporations, and of huge debts, huge dams and huge arms stockpiles, it becomes clear that poverty, hunger, malnutrition, and starvation may have far more to do with political, cultural, and socio-economic components of food shortages than with sheer numbers alone. This is not to understimate the immensity of the population problem or to minimize the difficulty of its solution.

*The Political Economy Perspective.* Barry Commoner (1974) faults socialist as much as capitalist economic theories for neglecting the biosphere as a major factor of production, but regards both poverty and growth of population as outcomes of colonial exploitation. The world, he believes, has enough food and resources to support nearly twice its present population. The problem, in his view, is a result of gross distributive imbalances between the rich and the poor, and requires a massive redistribution of wealth and resources to abolish poverty and raise standards of living in order to wipe out the root cause of overpopulation. The alternative to this humane solution is the unsavory one of genocide or natural destruction.

A study of environmental destruction in southern Honduras by Susan Stonich (1989) illustrates the power of a perspective that combines the concerns of political economy, ecology, and demography. Her conclusion is that environmental degradation arises from fundamental social structure and is intricately connected to problems of land tenure, unemployment, poverty, and demography. She identifies political and economic factors and export-promotion policies of international lending institutions and aid agencies as the key elements of a development policy for the whole of Central America that is likely to lead to destruction of the remaining tropical forests, worsen poverty and malnutrition, and increase inequality and conflicts within and between nations. Government policies encourage commercial agriculture for earning foreign exchange in the fact of mounting external debt, which rose by 170 per cent in just seven years to cover three-fourths of the 1986 gross national product. The expansion of export-oriented agriculture and the integration of resource-poor rural households into the capitalist sector, often by ruthless and violent means, concentrates highest population densities in the most marginal highland areas and encourages intensive land-use and adaptive strategies that acclerate ecological decline. Between 1952 and 1974, as a result of changes in land-use patterns, forest land declined by over two-fifths and the area in fallow by three-fifths. In the same period, food crop production was reduced drastically while the pasture area rose by more than half regionally and by over 150 per cent in the highlands, where the number of cattle rose by about 70 per cent. By 1974, a third of all rural families were landless; two-fifths were below the subsistence level in 1979.

The result has been the evolution of a class of rich peasants raising export-oriented cattle and cash crops, a class of land-poor and landless peasants and wage labourers, and a class of middlemen opperatives who serve as transportation links in an expanding regional and national network. The whole socio-economic structure has a most deleterious effect on the regional ecology and environment. These patterns are being repeated all over Africa and Asia. Even the "green revolution" provides only a temporary respite. Its recurrent and increasing high capital requirements for seed, fertilizer, insecticide, water, land, and machinery wipe out the small farmers and landless labourers. It destroys peasant agriculture, exposes the monocultures to destruction by disease and pests, magnifies inequality, and sows the seeds of social instability and rural strife. To those who subscribe to the political economy perspective, the biological explanation thus appears to be too simplistic. It overlooks the social context of development and land distribution within which worldwide destruction of traditional agriculture and the rain forests is now occurring.

In sum, these considerations bring out the fact that debates surrounding resource distributions and the control of population and consumption patterns are neither entirely scientific nor purely ecologically inspired. As Barry Commoner (1974) points out, they are political value positions. Will the changes come voluntarily, or will they involve totalitarian nightmares? "Sustainable development" and "traditional ecology" hold out two contrasting possibilities for the future.

At the heart of the 1987, Brundtland Report of the World Commission on Environment and Development is the idea of "sustainable development" that has become the rallying point for diverse agendas linking poverty, underdevelopment, and overpopulation to environmental degradation and "environmental security." The report defines sustainable development as "development which meets the needs of the present without compromising the ability of future generations to meet their own needs." Its popularity lies in its ability to accommodate the opposing idea of limits to growth within the context of economic expansion, but with a new twist. As pointed out by Gro Brundtland (1989), the "central pivot" of the notion of sustainable development remains "progress, growth, the generation of wealth, and the use of resources." The imposition of limits on consumption is then justified in order to protect the resource base of the environment both locally and globally. At the same time, continuous economic growth is held essential to meeting the needs of the world's neediest. In fact, the Brundtland Report indicates that "a five-to tenfold increase in world industrial output can be anticipated by the time world population stabilises sometimes in the next century."

Sustainable development is also seen as a strategy to enhances global security by reducing the threat posed by conflict and violence in an inequitable and resource-hungry world. To this end, it promotes a commitment to multilateralism, with a call for strong international institutions to ward off the new threats to security and for the collective management of global interdependence. As a result, the interests of economic growth and the environment are seen as mutually reinforcing rather than contradictory. The Brundtland Report (1987) duly notes that ecology and economy "are becoming ever more interwoven—locally, regionally, nationally, and globally—into a seamless net of causes and effects." The 1990 Worldwatch Institute Report predicts that the world will have a sustainable society by the year 2030. Meanwhile, the challenge, as Arnold states, is to ensure that the sustainability vision "is not trivialised or, worse, used as one more way to legitimise the exploitation of the weak and vulnerable in the name of global interest and solidarity."

To its credit, the Brundtland Report singled out some forms of economic growth that destroy resources and the environment. The present one-trillion-dollar expenditure on armaments, for example, constitutes "more than the total income of the poorest half of humanity." According to the 1990 United Nations estimates, military expenditures in developing countries, which account for 75 per cent of the arms trade, have multiplied by seven times since 1965 to almost two hundred billion dollars, compared with a doubling by the industrialised countries. In addition, burgeoning debt, adverse trade policies, and internal instability constitute the over whelming obstacles to sustained development. With Africa'a total debt approaching two hundred billion dollars (half of its overall grass national product and three to four times its annual income from exports), average debt repayments amount to more than half the export income. The debt burden forces the African nations to concentrate on monocrop export agriculture to the detriment of food crop production, and pushes hungry and landless farmers and nomads to marginal lands that they overgraze and over exploit in order to survive. However, with respect to fixing the responsibility for deforestation, the Brundtland Report appears to be of two minds. In asserting that to "most farmers, especially the poor ones, wood is a 'free good' until the last available tree is cut down," the report partly sides with the "tragedy of the commons" argument, accusing the poor farmers of being "both victims and agents of destruction." On the other hand, it also points to a different cause: "The fuelwood crisis and deforestation—although related—are not the same problems. Wood fuels destined for urban and industrial consumers do tend to come form the forests. But only a small proportion of that used by the rural poor comes from forests. Even in these cases, villagers rarely chop down trees; most collect dead branches or cut them from trees."

In any case, the Brundtland Report is now the focal point for global environmental efforts, even though in the United States it remains what the environmental activist Bettie Eisendrath has called "America's best-kept secret." It will undoubtedly play a crucial role in the United Nations conference on the global environment to be held in Brazil in 1992.

Patterns of Human social organisation and technology

use reflect the vision a people have of themselves and of their place in the universe. According to Karl Polanyi (1974), the question of how to organise human life in a machine society confronts us with a new urgency: "Behind the fading fabric of competitive capitalism there looms the portent of an industrial civilisation with its paralysing division of labour, standardisation of life, supremacy of mechanism over organism, and organisation over spontaneity. Science itself is haunted by insanity. This is the abiding concern". Bell (1976) has made the critical point that while the dominant nineteenth-century view of society as an interrelated web, a structured whole unified by some inner principle, still rules Marxist and functionalist thought, it is no longer applicable. On the contrary, society today is composed of three distinct realms—the techno-economic structure, the polity, and the culture—each obedient to a different axial (governing) principle, having differed rhythms of change, and each following different norms that legitimate different and even contradictory types of behaviour. The discordances between these realms are responsible for various contradictions within society. Bell has proposed the creation of a "public household" to overcome the disjunctions between the family, the economy, and the state through the use of modified market mechanisms to further social goals. At the same time, he is convinced that the crisis is a spiritual one of belief and meaning. "But today what is there left in the past to destroy, and who has the hope for a future to come? . . . What holds one to reality, if one's secular system of meanings proves to one an illusion?" he asks. His answer is the return in Western society of some conception of religion to restore the continuity of generations and provide a ground for humility and care for others. The task is truly formidable, for as he admits, "such a continuity cannot be manufactured, nor a cultural revolution engineered".

It is doubtful that the problems of order created by the "Normal" but dangerous disrelation between the life-sustaining (ecology-economy) and order-maintaining (sociopolitical) systems of contemporary society can be corrected by the creation of a miracle hybrid "public household," protected and nurtured by the polity and the household to serve the interests of the technoeconomic structure, and by the side-door entry of the "religious" to provide for the integrative and "higher" -order needs of a socially disjointed and spiritually vacuous society. Even the frantic use of a "holistic" ecological approach is bound to fail if its actual goal is somehow to dominate or hold on desperately to a sundered reality in which everything is so hopelessly *dis*related to everything else. The high-powered technoeconomic structure, driven by the insatiable demand for energy, resources, and markets, is inherently antiecological. Its immensity of scale and utilitarian thrust not only destroy traditional socio-economic structures but also set in motion irreversible and ecologically damaging global processes whose attempted solutions greatly magnify the problems. Marston Bates states that humans now truly constitute a new geological force. While the West is involved in a massive effort to take over the life-support systems of the earth, it is clear that a dependent part cannot grow infinitely at the expense of the others, or usurp the whole for its own purposes within conditions of environmental finitude.

A.K. Saran (1978) does not doubt in the least that the ecological crisis is a self-inflicted one, because and entropic environmental system and an infinitely expanding economy and technology are mutually incompatible. Furthermore, he points out that in the modern, historicist, horizontal framework of thought governed by a demiurgic anthropology, there can only be a technological solution to the problem of order in the socio-political realm. Ideally, it would require the destruction of this cosmos and the reconstruction of a new heaven, a new earth, "for if it were of a lesser order, a fundamental alteration of the nature of earth and sky could release forces and generate possibilities that must necessarily prove beyond the powers of the supposed masters of the earth to control.... For not having the power to reserve processes once started, (they) become their victims at the next stage".

Saran's main argument is that since the modern system excludes a truly vertical dimension, it has absolutely no authoritative foundation. It does not provide a coherent worldview or the proper regulative principle to satisfy the needs of the different orders in a unitive way. Where centricity and true hierarchy have been rejected, a piecemeal approach will be relied on the interest of the strong to deal with the consequences of a discordant and disharmonious order. In addition to generating tremendous violence, universal disorder, and planetary destruction in the desperate attempt to hold the parts together under its hegemony, such an approach is bound to fail. Since, the symbolic is not an integral part of the modern literal consciousness, the attempt to appropriate Mother Earth or other symbols, such as that by the proponents of Gaia, may be ideologically seductive but is both scientifically irrelevant and spiritually vacuous. Since, evolution has been the master concept to organise and rearrange the world in human terms, the ontology of modern science is necessarily anthropocentric. Saran's conclusion, therefore, is that there can be no ecological science unless it is grounded in traditional cosmology.

In a study of the Tukano Indians of the northwest Amazon, G. Reichel-Dolmatoff shows how aboriginal cosmologies, myths, and rituals . . . represent in all respects a set of ecological principles . . . that formulate a system of social and economic rules that have a highly adaptive value in the continuous endeavour to maintain a viable equilibrium between the resources of the environment and the demands of society.

The cosmological myths which express the Tukano world-view do not describe Man's Place in Nature in terms of dominion, or mastery over a subordinate environment, nor do they in any way express the notion of what some of us eight call a sense of "harmony with nature". Nature, in their view, is not a physical entity apart from man and, therefore, he cannot confront it or oppose it or harmonise with it as separate entity. Occasionally, man can unbalance it by his personal malfunctioning as a component, but he never stands apart from it. Man is taken to be a part of a set of supra-individual systems which—be they biological or cultural—transcend our individual lives and within which survival and maintenance of a certain quality of life are possible only if all other life forms too are allowed to evolve according to their specific needs, as stated in cosmological myths and traditions . . . . This cosmological model . . . constitutes a religious proposition which is ultimately connected with the social and economic organisation of the group. In this way, the general balance of energy flow becomes a religious objective in which native ecological concepts play a dominant organisational role. To understand the structure and functioning of the ecosystem becomes therefore a vital task to the Tukano.

However, modernity in its essence has been totally destructive of the traditional vision of human nature, our proper place in the "web of life," and our conception of the ultimate good. Polanyi points out how with the modern separation of "economy" as the realm of hunger and gain, our animal dependence upon food has been bared and the naked fear of starvation permitted to run loose. Our humiliating enslavement to the "material", which all human culture is designed to mitigate, was deliberately made more regorous. This is the root of the "sickness" of an acquisitive society that Tawney warned of . . . [T]he task of adapting life in such surroundings to the requirements of human existence must be resolved if man is to continue on earth.

The post-World War II creation of the global economy through the idea of "development" is the other half of the story. As pointed out by Wolfgang Sachs, and in line with Hawley's observation, the concept of development provided the United States with the vision of a new global order in which the former colonies were held together not through political domination but through economic interdependence. But [t]o define the economic exploitation of the land and its treasures as "development" was a heritage of the productivist arrogance of the 19th century. Through the trick of a biological metaphor, a simple economic activity turns into a natural and evolutionary process. [Soon] traditions, hierarchies, mental habits—the whole texture of societies—were all dissolved in the planner's mechanistic models . . . patterned on the American way of life.

However, even after nearly two decades of development work, the results were far from heartening. Instead of declining, inequality, poverty, unemployment, hunger, and squalor actually increased many fold in all "developing" countries.

To summarise: while the expansionist vision ties ecology, economy, and polity together, and the neo-Malthusians add biology to the list, it is in "sustainable development" that all these orders are firmly knit together—but at a price. The paradoxical nature of the term *sustainable development* arises from the fact that it attempts to combine the contradictory notions of limits to growth and active growth promotion. However, if the key to maintaining ecological integrity is economic self-sufficiency and production for use, then the problem today is surely one of the inhuman scale of enterprise based on the "techniques of degradation", which serve nothing higher than human self-interst, and of the concept of man as an economically rather than a spiritually determined nature.

Roy Rappaport (1976) has documented how the Maring of New Guinea support as many as two hundred people per square mile by cultivating nearly forty-five acres of cleared forest at a time, without damaging the environment. But then they look at the world through very different eyes!

## THE ENVIRONMENTAL MOVEMENT

Sociologists have paid significant attention to the study of the nature and dynamics of the environmental movement,which by the early 1980s involved over 3,000 organisations in the United States alone, though only about 250 operated on the national or multistate level.

The overriding emphasis of the environmental movement has been reformist. Concen with conservation and efficiency in resource and energy use, rather than with reallocating the production surplus among social classes, has dominated its agenda. Thus, the radical challenge to the dominant high-technology "treadmill of production," and the "softening" of the resource perspective of environmentalism by the emergence of equity concerns surrounding the new "appropriate technology" and "deep ecology" movements, have remained largely rhetorical, even as the focus on efficient resource use in production has made common cause between their membership and elites interests. Schnaiberg (1983), however, sees no upswelling of such redistributive politics in the near future.

One of the major research contributions has been the finding that public concern with the environment follows the same issue-attention cycle that is characteristic of most social movements. Thus, a review of current research by Dunlap (1986) revealed that public concern with environmental quality, which rose sharply in the latter half of the 1960s but peaked by 1970, declined throughout the 1970s, then rose again in force in the 1980s. As a result of the second Earth Day (1990), and the intense national and global publicity and debate on environmental issues, such as global warming and the unprecedented "environmental terrorism" in the Gulf region, public concern is at an all-time high in this United Nations "Decade of the Environment."

Environmental sociologists have explored the charge of "elitism" levelled against the environmental movement. A review by Morrison and Dunlap (1986) points out that the class composition of the environmentalist is similar to that of participants in other mass movements, that their idelogical commitments range from the cause of the privileged to the problems of the poor, and the reforms do appear to have had a somewhat disproportionate negative impact at the lower end of the socio-economic scale. Research in this and other areas is summarised in Buttel (1987).

Overall, while the environmental movement has had a significant world wise impact on the way people and politicians look at the environment and its problems, this change has not yet been translated into comprehensive and effective policies to promote and protect environmental interests. Internationally, the United Nations' 1972 Stockholm Conference marks a watershed in the history of global action to protect the environment. It led to the formation of the United Nations Environment Programmes and to various important declarations, protocols, and treaties on deforestation, desertification, management of oceanic and other common resources, transboundary pollution, global climatic change, international legal and regulatory mechanisms to ensure compliance, and the like (Leonard 1990). A concise and comprehensive history of the birth and growth, and the philosophy, of the global environmental movement is given in Singh (1985). Sinh also has outlined the agenda to harness the tremendous potential power of people's elected representatives in the service of the environment. The 1992 global conference on the environment to be held in Brazil will review two decades of progress since Stockholm, and is likely to bring the deeply divergent views of the North and the South on the environment into sharper focus.

The newly emerging concern with human ecology and the environment has moved into the mainstream of public life as a major national and international concern. This provides important opportunities for environmental sociologists to contribute to the understanding and solution of these problems. Constance Holden (1989) has highlighted the report of the National Academy of Sciences that outlines an agenda for both micro and macro social scientific studies of "anthropogenic" stresses on the resources and the environment in the north circumpolar region, which has general application. The fragile arctic region has a great wealth of natural resource. It comprises one-tenth of the global area and has eight million people, of whom a quarter are natives. The report placed major emphasis on interdisciplinary studies, particularly those linking the social and physical sciences and basic and applied research. It emphasised the need for drawing on native knowledge and put urgent priority on issues such as cultural survival and the allocation of scarce resources. It also asked the social scientists to come up with models generalisable to other areas.

An interesting insight concerns how each of the several identities of the Arctic (e.g., as homeland for the natives, as a "colony" exploited for its natural resources, and as the last wilderness) results in a distinctive approach to the human-environment relationships. "These aproaches have come increasingly into conflict as subsistence hunters and commercial interests vie for limited stocks of fish and game; communities are shaken by boom and bust cycles in scrambles for mineral resources; and rapid modernisation has inflicted trauma on native cultures". The committee identified three areas of interest to the social scientist. In the area of human-environment relationships, there is need for studies on conflict resolution to strike a balance between commercial needs and the interest of subsistence hunters, sportsmen, and conservation. The second area pertains to community viability, for which a systematic approach is needed to help develop physical and social services infrastructure to meet the special climatic needs of the region. A final area pertains to the study of the impact of rapid social change (single-industry cash economy, the snow mobile revolution), which is exacting a heavy price from the local inhabitants in terms of higher rates of alcoholism, suicide, stress, loneliness, accidents, and violence.

The commitment to protect growth or a certain way of life has led conventional human ecologists and the neo-Malthusians to disregard the minimum well-being or sheer survival of the rest of humankind. In fact, Hawley admits that while "competition," resulting from demand exceeding the carrying capacity, may account for the exclusion of some contestants from access to their share of a limited resource, it does not shed "any light on what happens to the excluded members of a population after their exclusion." This serious neglect of the concern for the underdog and the undermined is matched by the self-admitted tendency of the human ecologists "not [to] confront policy matters directly".

Edgar Borgatta (1989) has sought to develop an important

field called "proactive sociology, with a view to closing the wide gap between sociological theory and practice and to save sociology from sheer irrelevance. Sociological approaches, even when application-oriented, have been largely timied, in active, or merely reactive. Rather than waiting to study only the after effects of "all in the path"—the Three Mile Island radiation leak or the *Exxon Valdez* oil spill—or stepping in at the end of the "issue-attention" cycle, when the problem is historically interesting but socially irrelevant, a proactive sociology will concern itself with the dynamics out of which problems arise, anticipating potential problem areas and their alternative solutions as the means to translate desired values into effective policy. This will involve identifying possible futures and the consequences of action or in- action for their attainment, a policy dimension ignored by sociologists, despite their belief that this may make all the difference in a fast-changing and turbulent world in which the ability to handle and manage change requires the ability to anticipate change and to adapt social structures to changing requirements. To this end, the sociologists needs to task if what he or she is doing will make an impact and be useful to society. The fundamental assumption here is that if we know something about the impact of social structure on behaviour, we should be able to propose models for changes in social structures that will effectively implement values which have priority status in society Sociologists are then obligated to "address societal values more directly by providing alternate models of potential changes and exploring the consequence these changes may produce if identified values are implemented"

Following this lead, a "proactive environmental sociology" would broaden the scope of the "sociology of environmental issues" by focusing specifically on the changes that are required to effectively implement stated values (equity, social justice, "future generations," conservation, sharing the global commons, clean and healthy environment, changes in life-styles and consumption patterns, and the like), and by exploring the possible consequences of these changes. Thus, David Mahar, an adviser to the World Bank, has argued that blaming peasant colonists for deforestation is "tantamount to blaming the victim" for "misguided public policies" that promote road building, official colonisation of the forest, and extensive livestock development, and that " purposely or inadvertently encourage rapid depletion of the forest." This definition of the situation led Mahar to propose an "alternative development model" that would put government action on hold so that, based on land-use surveys, "(l)ands found to have limited agricultural potential—virtually whole of *terra firme* [sic] of Amazonia—would be held in perpetuity as forest reserves...." These and other unconventional conclusions are stated by Mahar as his own and carry the disclaimer that they do not necessarily represent the views and policies of the World Bank itself.

This example also brings out Borgatta's point that a proactive stance may involve the espousing of unpopular positions. It may lead a proactive environmental sociologist to examine the role of established institutions and values (crass individualism and the impact of "anthropocentric," "cowboy," "superpower," and "sustained development" approaches to the use of finite resources and a fragile environment, corporate nonaccountability and the global impact of multinationals, the state, and the like) in order to facilitate the creation of environmentally sustainable social structures that implement stated values. On this basis, a systematic concern with the application of knowledge would lead to a "proactive environmental sociology" that would prompt the sociologist to formulate alternative policies with respect to the set of environmental values or goals that are to be implemented. This will also ensure that the applied aspects of "environmental sociology" will flourish within the discipline and not become detached from sociology, as has been the fate of industrial sociology and many other areas in the past.

Environmental sociologists have complained of the lack of the unifying focus within the field, and have noted its specialised, fragmented, and dualistic tendencies, which hinder concept and theory development. This should be a cause for serious concern insofar as the new human ecology is supposed to provide a holistic, integrated understanding of human—environment interaction. In addition to the problems, surrounding functionalist as well as Marxist categories and assumptions is the difficulty of adapting bioecological concepts to the human context. Notions such as ecosystem, niche, succession, climax communities, balance of nature, even evolution, have no clear social referents and pose formidable problems of inappropriate or illegitimate transferral of concepts. Thus, while one finds constant reference to urban or social "eco-systems" in the literature, the wide-ranging, even global, energy-exchange patterns make the boundaries so diffuse that it becomes impossible to locate an urban *ecosystem* in time and space, at least in biological terms. Or, if humans are defined as niche dwellers, the term *niche,* "if adopted directly from biology would produce only one worldwide niche for the entire global species, a result that would render the concept useless. How can the species problem be overcome in adapting such a concept to human ecology?"

Terms such as *the environment* are not easy to define or conceptualise; nor are ecological chain reactions, multiple causal paths, and feedback mechanisms in complex ecosystems easy to delineate. In recognition of the substantial difference between human and bioecological orders, some human ecologists, such as Hawley, have moved away from bioecological models. Thus, Hawley is highly critical of the neo-Malthusian application of the "carrying capacity" notion, on the ground the "while the argument may be suitable for plants and animals, its transfer to the human species is highly questionable." While still shying away from assigning a critical role to human agency, or even a policy making role to the human ecologist, Hawley has nonetheless broadened the scope of his theory by incorporating culture and norms as ecosystem variables. Rappaport has raised more basic objections: To treat the components of the environment as if they were mere resources is to view them exclusively in economic terms and invite "the use and abuse of biological systems of the classes and the neglect of moral and aesthetic considerations in general. Whatever may be meant by the phrase 'quality of life', exploitation does not enhances it"

Human ecologists, in general, have not dealt adequately with such concerns. nor with the problems of power, domination, and the role of the state and of "values" in human-environmental relationships. At a minimum, one needs to know the role of the state in the regulation, maintenance, expansion, suppression, and "resocialisation" of people and societies. If ecosystems are constituted of interdependent parts, one needs to known the nature of the reciprocal relationships among the parts and among the parts and the ecosystem. Rappaport (1978) has drawn attention to the maladaptive tendency of subsystems to become increasingly powerful and to dominate and use the larger system for their own benefit, to the detriment of the general interest and the adaptive flexibility of the system. He mentions the dominating positions occupied by huge corporations and the "military industrial complex" as examples. More broadly, Rappaport ties pollution, "resource" depletion, and the diminution of the quality of life and the destruction of its meanings to the scale of modern societies and the voracious appetites of their industrial metabolisms. Thus, while he does not deny that population increase may have a negative impact on the quality of life, he has little doubt that the real cause of ecosystem destruction and the deterioration of the quality of life is to be found in the way societies are organised, not in their population trends. If that is so, what alternative do humans have?

Within the human ecological perspectives, environmental problems are seen as arising either from the unplanned nature of growth and expansionism, and its attendant externalities and "commons" tragedies, and growth and market restrictions (the pro-growth, expansionist perspective), or as a result of the excess of population over the carrying capacity of the environment (the neo-Malthusian perspective). To restore ecological balance and environmental health, human ecologists place their faith in value consensus, rational planing, systems theory, computer models, economic growth, trickle-downs, market mechanisms, technological fixes (the pro-growth perspective) or in limits to growth, sustainable development, sticks and carrots, benign neglect, the triage, die-offs, outright compulsion, even genocide (the neo-Malthusian perspective). Within the political economy perspective, on the other hand, the emphasis is on internal contradictions, uneven development, centre-periphery relations, capitalist exploitation, the role of multinationals and the state, trade imbalances, and the treadmill of production. To ensure environmental protection, distribution, social justice, and quality, the proposals from a political economy standpoint range from social revolution, conflict, and confrontation to social welfarism and mixed economies.

The political economy perspective is critical of the basic assumption of the Chicago ecologists that changes in pollution, organisation, and the technologies of movement explain expansionary movements and territorial arrangements. By allowing planners to alter spatial forms to dissipate class conflict and social unrest, Smith believes, the perspective becomes a powerful depoliticising weapon in their hands. He favours "client-centred" planning, which does not assume that "physical structure determines social structure," but holds that both are shaped by the economic and political structures of society which provide selective access to opportunities, and further discriminatory patterns of land use and investment. Smith, therefore, offers. "conflictual planning" on behalf of the poor and the powerless, to call attention to the hidden social costs of development and to increase the political costs of pursuing repressive policies disguised as rationally planned allocational, locational, and investment choices. This approach poses three basic questions: "Whose values, interests, and social actions will determine the purpose, pace, and direction of historical change? Can the costs and benefits of historical change be distributed fairly? Can the changes that do occur further the cause of social justice?".

Schnaiberg (1980) has identified three responses to the contradiction between production expansion and ecological limits: (1) the expansionist, which will be temporary, increasingly unequal, environmentally stressful, and authoritarian; (2) the business-as-usual, which will be unstable, socially regressive, unequal, and of limited environmental value; (3) the ecological, involving appropriate technology, reduced consumption, and reduced inequality, which will be the most durable but also the most socially disruptive and the least desirable. William Ophuls (1977) has called for a rejection of Alexander

Hamilton's ruling vision of power in order to realise the Jeffersonian vision of republican simplicity. Schnaiberg's own preference, short of a social revolution, is for a mixed social democratic system like Sweden's, with some production expansion and improved welfare distribution under close state supervision.

However, this solution does not quite address the critical concerns of the environment or the needs of three-fourths of humanity. It presents to the world the antiecological model of the "treadmill of production" under a more benign form. The common problem, as Hilary French (1990) points out in a Worldwatch paper on environmental reconstruction in socialist nations, is that of finding the proper balance between sufficiency and excess, which he says will be as difficult for the socialist countries as it has been for the West. He is afraid that under the impact of powerful forces, the socialist countries may end up repeating some of the West's errors. In this context, he points out how Czechoslovakia's president, Vaclav Havel, has identified "the omnipresent dictatorship of consumption, production, advertising, commerce, and consumer culture" as the common enemy. Nancy Anderson, the well-known New England environmental educator and activist, blames un-controlled greed for the frightening global environmental degradation and for overpowering our sense of responsibility to future generations.

Gandhi, aware both of the fatal attraction and the destructive potential of wanton materialism, saw it as constituting the gravest threat to human freedom, survival, and environmental security. He, therefore opted for a simple, nonexploitative, and ecologically sustainable social order. Such a decentralised social order, based on truth and non-violence, is to be governed by the metaphysical determined optimum levels of wants, technology, and resource use fitted to the requirements of the human scale. In the interim, he demanded that the rich become trustees of the poor in order to serve justice, to mitigate the negative impacts of the differentials of wealth and power, and to avoid class conflicts. His radical vision of a normal social order, nowhere realised as yet, provides a useful yardstic for measuring how ecologically sound and environmentally sustainable a society is in its actual operation. Noting that the world has enough for everyone's needs but not for everyone's greed, Gandhi was convinced that such a social order would come about only if the means of production of the elementary necessaries of life remain in the control of the masses. These should be freely available to all as God's air and water are or ought to be they should not be made a vehicle of traffic for the exploitation of others. Their monopolisation by any country, nation or group of persons would be unjust. The neglect of this simple principle is the cause of the destitution that we withness today not only in this unhappy land but in other parts of the world too.

From this point of view, while there is little disagreement that overpopulation aggravates environmental and other problems, the attempts to eradicate the root causes of social instability, inequality, and poverty are bound to be far more effective in the long run than the impressive but partially effective approaches to population control. Brian Tokar (1988) has pointed out that, historically, rapid increases in population occur when people become dislocated from their traditional land base and become less secure about their personal and family survival. On the other hand, populations become stable when the future is secure, the infant mortality rate is low, social choices for women are expanding, and parents are not worried about who will support them in their old age.

How to effect the radical changes required to restore the proper ecological balance and preserve the biocultural intergrity and diversity of the global "household", but "without the most fantastic 'bust' of all time", is the formidable challenge and the urgent task facing humankind. This will involve a redirection of the vast, creative human energies away from a self-defeating and ecodestructive expanionist and wasteful orientation, and their rechanneling into the life-giving and life-promoting forms of human action and human social organisation.

# PART I
# ETHICS OF ECOLOGY

# 2

# HISTORICAL CONCEPTS OF ECOLOGICAL BALANCE

Ecological balance is a kind of stability or persistence that is attained through counterposed forces, and may be conceived of as either precarious, in the analogy of a tight-rope walker, or resilient, in the analogy of a well-balanced character. A structure can be stable or persistent without being balanced—for example, a mesolithic stone circle. Ecological balance involves the relations and interdependencies between living things, and their (non-living) environments. It may be understood in a functional, structural, or a dynamic sense. Specifically, it may be a feature of ecosystem function, population dynamics, or community structure, and different models of balance—for example, hydraulic, organicist—are appropriate in the different cases. Ecological balance is subject to relativities of both time and space, in the sense that a process or structure that is balanced in relation to a particular period of time or a particular region of space may not be so in relation to some other period or region, and *vice versa*. The time scales involved may vary from short-term cultural time, through medium-term ecological time, to long-term evolutionary time.

The spatial region involved may vary from that of

pond to that of the whole planet. The mere fact that matter or energy is conserved in any given system is not sufficient to found the notion of ecological balance, which clearly requires the repetition or persistence of features at an ecologically meaningful level. Among the more obvious apparent signs of balance in terrestrial nature are the daily and seasonal cycles, and the fixity and fittingness (or adaptation) of species. Opinions differ over the degree and kind of balance that characterises the natural world. One view is that balance is an inherent feature of the natural world. Another view is that it is a contingent feature of the natural world. A third view might be that, like the curate's egg; it is balanced in parts. A fourth, moderately skeptical, view is that it is not a feature of the natural world at all, although it might have been. A fifth, radically skeptical, view would be to question whether it is even meaningful to ascribe balance to the natural world. It is a further epistemological question whether, given the great complexity of the natural world, we could ever get to know which, if any, of these views is true.

## 1. PRE-DARWINIAN CONCEPTIONS OF BALANCE

### (a) The Ancient Period

The conception of nature as balanced long precedes modern ecological analysis. Often, this conception has had a "religious" basis—on a broad understanding of that term; thus, many "worldviews" have regarded the natural world as providentially ordered. But the notion that such ordering is the work of a supernatural agency, rather than built into the fabric of things, is arguably the exception rather than the rural. Both Buddhist and Taoist philosophies, for example, like Western Stoicism, feature the idea of nature rightly ordered, not by some external governance but from the inside, and accordingly recommend a life "following nature," Popular religion however, perhaps answering the needs of common people who "felt the draft" rather more than the priviledged, was less convinced. The anarchic world of Homer's gods, for example, provided little reassurance that the world was rightly ordered. The legend of Pandora's box, which, when opened, released a host of evils upon the world, leaving hope as the only solace, gives characteristic expression to this outlook.

Early Western cosmologists construed nature as composed either (i) of a single element, such as water or fire, capable of many transformations; or (ii) of a finite number of different elements, such as earth, air, fire, and water interacting with or being tranformed into one another; or (iii) of an infinite number of elements finite in kind, such as atoms and void. The majority, while aware of fluctuating forces and elements—conflagrations in the case of Heraclitus, love and strife in the case of Empedocles—pictured such changes as staying within fixed limits. A notable exception were the atomists, whose almost "modern" outlook is most completely described in the work of the Roman author, Lucretius, *On The Nature of Things* (1st century B.C.). They postulated a real infinity of time and space in which temporary pockets of stability, in the form of clusters of vibrating atoms, occurred at random—literally, on account of a random "swerve"—in the incessant rain of atoms.

The current denizens of earth are merely the survivors from trial and error processes that the earth "mother', in recent times thought of as failing in her powers, has yielded up. While sharing with Stoicism a belief in the value of attaining an internal psychological balance—ataraxia, or tranquillity—the Epicurean atomists believed this was achieved by understanding rather than by following nature.

As regards living nature, the myth of Epimetheus recounted in Plato's *Protagoras* sets the prevailing tone. The Titan, Epimetheus, charged with distributing to earth's creatures their proper characteristics, gave to each the means of sustenance, and the means to protect themselves against the elements and against each other, "with the view of preventing any race from becoming extinct." The Greek historian Herodotus suggests that practices of cooperation between species, and differential rates of reproduction, have the same effect; while Aristotle contributes the thought that physiological limitations built into a creature's form constrain its ecological impact. What Egerton has termed "providential ecology" is epitomised in the Stoic picture of a stable and coherent world, held together and suffused by divine intelligence, and possessing the integrity of a living organism, so that, in Cicero's words, "nothing better adapted to endure could be imagined," Although, it was a system arranged "for the safety and preservation of all," in which every thing had its place and nothing was superfluous, human convenience was frequently emphasised. Chrysippus, a leading Stoic, notoriously suggested that pigs are given souls in order to keep their meat fresh, and that the role of fleas and mice, respectively, is to encourage cleanliness and tidiness in humans.

## (b) The Early Modern Period

Belief in providence, hierarchy, and order was readily assimilated into the medieval worldview. But views of the natural world tended to oscillate thereafter between the opposing poles of turn oil and tranquility, often appearing the reflect contemporary social concerns. At the same time, these contrasting views were seldom found in purely unmixed form. In the seventeenth century, darker visions surfaced in the likes of Thomas Hobbes' depiction of the state of nature as "nasty, brutish and short," and John Milton's "dalliance" with Satan in *Paradise Lost.* The belief in a providential ecology, associated with natural theology, and expressed in the metaphor of a great "chain of being," however, experienced a revival in the seventeenth and eighteenth centuries. Richard Pulteney spoke of "that perfect order and just subordination of all the several parts of nature," a point illustrated by William Derham, who remarked, how the balance of the animal world was "kept even" by the trade-off between rates of reproduction and life expectancy. In citing differential longevity as a cause of the balance he was anticipated by Thomas Browne, who added the suggestion that hibernation served to keep fierce animals out of mischief. Gilbert White, meanwhile, found evidence aplenty of cooperative accomodation, noting how dung from the cattle in his local ponds harboured insects that provided food for the fish. (This spirit of collaboration, however, did not extend to noxious insects, concerning which he declared that a compendium of known and likely means of destroying them would be a most "useful and important work.)" These harmonious relations, which served to maintain not must species, but also permanent forms of association, were in turn codified by Linnaeus, whose analysis of the natural world suggested that propagation, preservation, and death were the chief factors maintaining the balance. Under this view, apparent exceptions to ecological balance did not go unnoticed, but they were easily explained away: plagues were sent to chastise or test us; apparently extinct species were still to be found in so-far unexplored regions. Only when extinction was taken seriously would these explanations of plagues and pests cease to be acceptable. New sources of tension emerged with the early nineteenth-century debate over the formation of the earth's surface, between catastrophists led by Georges Cuvier, who believed that the present state of the earth's surface could only be explained on the supposition that it had undergone periodic convulsions, and the uniformitarians led by Charles Lyell, who believed that the phenomena could be adequately explained by familiar and more gradual processes. A twist to this debate was that the catastrophists position, rather than the uniformitarian, was thought to be more consonant with religious belief and supernatural agency.

## 2. DARWIN'S "CONSIDERABLE REVOLUTION"

### (a) The Theory

Against the background warnings of geological instability that had been sounded by Hutton, Cuvier, Humboldt, Lyell, and others, and whether the cause of change was catastrophic or uniform, Engels in his *Dialectics of Nature* was later to comment on the striking anomaly of postulating fixed organic species in a changing inorganic environment. When Darwin offered his theory of natural selection and a credible explanation of species mutation, he modestly claimed to "dimly foresee" how it might involve a "considerable revolution" in natural history. Essentially the theory of natural selection shows how the hereditary characteristics of a population of organisms may change over time due to their differential reproduction arising out of the "struggle for life." Its effect was to replace the conception of the natural world as an enduring, and usually closed, system with a conception of its as an open-ended historical process. A further effect was to render the notion of balance problematic. One view is that the theory of natural selection undermines the notion of ecological balance completly. But another view is that it simply serves to explain it. Ecological theory since Darwin, while adding enormously to our understanding of the constraints and contingencies to which natural phenomena are subject, has so far failed to resolve the issue.

### (b) The Implications of the Theory

In favour of the view that natural selection undermines the notion of balance is the emphasis in the theory on struggle—the "dreadful but quiet war," as Darwin called it (*Notebooks,* 1838)— with its implication that no particular outcome is assured. Under the hypothesis of natural selection the only constant is change, while responses to change are provisional and makeshift. Stability might be compatible with change if the change exhibits recurring pattern or direction, but natural selection assures us of neither. In favour of the contrary view that natural selection explains balance, it might be observed that the scientific worldview had already assimilated corpuscular, that is, atomic, theory and therefore the idea of wholesale motion at the microscopic level, with out this apparently destroying the belief in a well-governed universe. Furthermore, evolutionary change was sufficiently long drawn out to leave the relatively stable world of ordinary experience intact. Even the subsequent idea that the course of evolution has been "punctuated" by periods of rapid change, given that these are externally caused, need not disturb the belief in the overall stability of the system, in light of its demonstrable powers of recovery. Moreover, Darwinian theory does not supplant, but indeed explains, the phenomenon of adaptation, because it yields the idea that, through the mechanism of natural selection, populations of organisms at least have the capacity to track the changes that are constantly occurring around them.

## 3. THE CONCEPT OF BALANCE IN ECOLOGICAL THEORY

Whether, as some might have hoped, the new "science" of ecology that Darwin's work ushered in has succeeded in placing the question of balance on a sounder, or more testable, footing, still remains unclear. But at least there have been considerable conceptual advances made, and a resulting clarification of the issues. Three types of balance have been distinguished: constancy, or the extent to which some population, community, or system remains unchanged; inertia, or the capacity to resist disturbance; and resilience, or the capacity to respond to disturbance. The two latter notions of balance are "static" inasmuch as they refer to a propensity to retain or revert to some initial state. One thing that can immediately be inferred from thus analysis is that the issue of balance may not be an all-or-nothing matter: something that is constant might not be resilient, and so forth. Moreover, the presence of inertia or resilience is clearly relative to the nature of the disturbance involved: resilience in the face of disturbance of kind X—for example, fire following a lightning strike—will not simply resilience in the face of disturbance of

**Table 2.1**

**Types of Resilience**

| *Type* | *Characterisation* |
|---|---|
| (i) Elasticity | The speed of recovery to some initial condition, following diturbance |
| (ii) Amplitude | The extent of the disturbance from which recovery to the initial condition will occur |
| (iii) Hysteresis | The extent to which the pattern of recovery mirrors the pattern of disturbance. |
| (iv) Malleability | The degree to which the state following recovery simulates the initial state |

kind Y—for example, a flash flood. In his classic account of intertia and resilience, W.E. Westman *(Bioscience* 28, 1978), following G.H. Orians, has distinguished four kinds of resilience; see Table 2.1

Ecologists usually discuss the issue of "balance", or near equivalents such as "stability" and "equilibrium," in connection with segments of the natural world identified as "communities" or "ecosystems." It should be noted, however, that some strongly prefer to use a more noncommittal term such as "patch" to refer to these segments, on the grounds that the use of terms such as community and ecosystem are likely to *predispose* one to find structure. In a given community, among the commonly recognised signs of apparently nonaccidental balance or stability are: constancy of its constituent populations, recurring patterns of abundance, correlations between abundance and size of organisms, patterns of feeding relations, and so forth. Two questions arise: whether this apparent stability is real; and what might account for it. It is natural to think that relations between the constituent organisms—"biotic interactions"—will play a large part in determining the structure, composition, and dynamics of a community. One early attempt to explain community structure focused on the feeding relation.

## (a) Trophic Relations, Trophic Structure

Although Charles Elton notoriously remarked that "the balance of nature does not exist," his work on animal communities was nevertheless seminal in its approach to community structure. Elton held that all animal communities are structured in a particular way and that the crucial factor determining this structure is FOOD *(Animal Ecology,* 1927). He distinguishes four general principles that regulate animal communities, namely food size, the food chain, the niche, and the pyramid of numbers:

(i) *Food size.* Athough, there are exceptions, Elton was struck by the fact that there are upper and lower limits to the size of food that any particular animal can eat (he cites the difficulty that a human would find trying to make a meal of bilberries). This is a function of physiological and anatomical features such as bite size and the degree of nourishment afforded by any particular source of food in relation to the effort, size, and talent needed to acquire it.

(ii) *Food chain.* Facts about what an animal can and cannot eat broadly determine a chain of predator-prey relations. More realistically, the "chain" is in fact a web, because any given species will probably eat and be eaten by a number of different species.

(iii) *Niche.* Narrowly conceived, an animal's niche is precisely its place in the food chain, as determined by the kind of animal it is (so distinguishing it form Darwin's concept of place that is occupied by individual animals). More broadly, it is the animal's status, role, or "occupation" in the biotic community of which it forms a part, and how it relates to its food and to its enemies.

(vi) *Pyramid of numbers.* The most abundant animals will be those at the bottom of the food chain who constitute food for the rest, and who themselves consume plants, detritus, and other nonanimal material. In the middle of the pyramid will be animal that both eat and are eaten by other species. The top predator is simply the species which is not numerous enough to constitute a food supply for any other species.

Elton held that these four principles ensure certain "constancies" in the structure of animal communities, which invariably are found to comprise three main classes—

of "producers," "consumers," and "decomposers."Also, remarkably similar niches—scavenging, sap-sucking, and so forth—are found in widely differing animal communities both on land and sea.

Although, Elton's work continues to attract interest—for example, the intriguing problem of why the number of trophic levels rarely exceeds four or five—much of its has been superseded. Perhaps, the chief limitation in Elton's approach concerns his exclusive focus on tropic relations. Many other kinds of biotic interaction, such as competition for resources, interference, mutual support, and converging or competing habitat requirements are now thought to play a crucial role in structuring community relations and therefore, possibly, in explaining community stability. Moreover, the trophic analysis itself has been found to obscure both (i) forms of organisation within trophic levels, and (ii) the subtlety of relationships between trophic levels—involving the number, type, and strength of the links. There is growing evidence that both these factors of food-web structure affect the dynamics and stability of the community at large. Finally, the niche concept has proved neither sufficiently fine-grained nor sufficiently well-defined in operational terms. Thus, although the theory suggest that niche overlap should be minimal, such overlap is common when resources are abundant. Moreover, species assigned to different niches on grounds of spatial or temporal separation may nevertheless be in competition if their (common) food source is excessively mobile, or if it is consumed at different times or at different stages in its life cycle.

### (b) Fluctuating Populations

When Elton denied the reality of the balance of nature he cited, in particular, irregularity in the fluctuations of animal populations both as regards frequency and degree.

But this seems to have more bearing on the presence of constancy than of resilience. Indeed, his further point, that variations in the numbers of one species have repercussions on the numbers of others, might be thought to tell as much in favour of balance—or at least of a balancing mechanism—as against it. Whether there is a mechanism regulating animal numbers and if so what it is, is a matter of contention.

Competing accounts can be divided into those that cite an external cause and those that cite an internal one. Possible external controls that are biotic include insect parasitism, disease, and shortage of food; possible abiotic controls include climate and lack of suitable habitat. Among internal controls to have been suggested are genetic (self) regulation—the idea that an increasing number of less viable genotypes accompany rapid increase in population (G. Chitty), and territorial (self) regulation—the idea that among some species there is an instinctive territorial sense that permits a certain number but no more to occupy a given habitat (V. Wynne-Edwards). A question raised about both hypotheses, however, is how far they are compatible with the theory of natural selection. How far such fluctuations are a function of the density of the relevant populations (i.e., how far they are "density dependent" or "density independent") is also an issue. What seems to be true of most, if not all of these hypotheses, is that they are capable of explaining the pattern of fluctuations that are known to have occurred. But the processes that they posit do not carry any guarantee about what the pattern of fluctuation is going to be in the future; they carry no built-in guarantee of balance. Moreover, they do not and cannot guarantee that numbers will not drop below the level at which a species is viable: clearly, they do not ensure against extinction.

One way in which excessive fluctuation might actually,

and paradoxically, contribute to constancy has been recently proposed by S. Pimm and J. Lawton. The usual explanation for the limit to the number of trophic levels is that the available community "energy" is not sufficient to support another level. If that were true, Pimm and Lawton argue, one might expect that communities in which primary productivity (i.e., the amount of energy captured by the "producers") was greatest would support more levels—which is not generally true. They suggest instead that top predator populations are simply not stable enough to support a further trophic level.

## (c) The Ecosystem

In 1935, Arthur Tansley proposed a new concept, the "ecosystem," which provided a unified framework within which to study both plant and animal communities together, their interactions with inorganic nature, and their interrelations with human communities also. Since, Clausius formulated the Second Law of Thermodynamics in 1850, it had been understood that in a general trend toward disorder the tendency was for energy to become dispersed. By capturing some of the sun's energy and making it available to animals, plants temporarily reverse this process. Set within Tansley's newly devised concept of the ecosystem, food could be viewed functionally, as partly comprising the means by which both energy and nutrients are transmitted through the biosphere. It was R. Lindeman's landmark paper on "The Trophic Dynamic aspects of Ecology" *(Ecology* 23, 1942) that paved the way for this more comprehensive level of analysis. He sets out what is essentially an "economic" analysis of energy transfer, measuring the productivity and efficiency of each trophic level, thus building a picture of what happens to captured solar energy as it moves through the ecosystem. Besides, the transfer of energy, the movement through the system of minerals and nutrients such as carbon, oxygen, nitrogen, and sulphur could be studied in the same way. The analysis of systems in terms of energy enables new facets of community structure and ecosystem function to be revealed. For example, there is recent evidence to suggest that both the rate of energy flow and the efficiency with which it is transferred (aspects of the type and strength of trophic links) contribute to stability.

During the middle part of the century, "ecosystem theory" was developed and formalised, notably by the brothers, Howard and Eugene Odum. Nature was conceived as a series of hierarchically ordered ecosystems a⁺ various stages of development. Distinguishing features of mature as opposed to immature ecosystems were drawn up, such as greater stability, increased diversity, and minimal loss of minerals and nutrients. These energy, mineral, and nutrient cycles, and the circumstances that facilitate and obstruct their operation, were open to quantitative measurement and became the subject of extensive study. Their existence served to reinforce the conception of nature as systematic and to that extent "balanced." Moreover, the discovery of mineral and nutrient cycles operating on a planetary scale, together with the moderation effects of life on oxygen levels and circumambient temperature, have encouraged some to believe that life has the capacity for self-regulation and self-organisation at a higher level. The idea has been given eloquent expression in J. Lovelock's "Gaia" hypothesis, which posits that the plant itself is just such a self-regulating system. The salinity of the sea, and oxygen and temperature levels, are far more constant, it is argued, than they would be if there were no life; and furthermore, this (relative) constancy is essential for the continuation of life. A more cautious reading of the same phenomena, however, would argue that the mineral and nutrient cycles, and the inevitable effect of the existence of life upon oxygen and temperature levels, simply explain how it is "as if" the planet is regulating itself—much in the way that Darwin's theory explains how it is "as if" nature were designed.

## (d) Ecological Succession

### *(i) The Theory*

Communities (ecosystems—or patches) appear to undergo repeatable directional and convergent changes of structure. This is the phenomenon known as "succession." As with evolution, so with succession, it is important to distinguish between the process and the theories advanced to explain the process. The "classic" theory of succession was anticipated by Thoreau and developed by Eugenius Warming, Henry Cowles, and Frederick Clements. It is nearly always advanced as a theory of plant succession, although, as Rory Putman observes, there is no obvious reason why its scope should be limited to plants—given the interdependence of plant and animal populations. In brief, the theory maintains that the character of the vegetation in any terrain is fundamentally determined by climate (and to a lesser extent, by geophysical factors). Following any process that yields a site relatively free of vegetation—whether natural such as fire, volcanic activity, or the retreat of a glacier, or caused by human agency—the site will first be colonised by "pioneering" species. According to Clements's version of the theory, these will affect the site in such a way as to render it suitable for new species, and facilitate their establishment.

In some cases, the new species will displace the original occupants. The process will continue until a relatively stable or climax state is reached that is most favoured by the prevailing climatic conditions. In practice, the process may be affected by a variety of contingencies;

but these are both anticipated and provided for by the theory. For example, the process may be interrupted, diverted or arrested. If interrupted, the succession is repeated. If diverted or arrested, it will, predictably, settle into a relatively stable form of substitute climax—called in these cases a "proclimax" and a "plagioclimax," respectively. It is by virtue of postulating that plant communities, although changing, are changing according to ascertainable laws, and that deviations are fully explicable, that the theory tends to support the conception of nature as balanced; furthermore, these communities are conceived as tending towards a steady towards a steady state.

In Clements's hands the theory was overlaid with a metaphysical view which saw plant communities formed in this way as "super organisms." He describes them as similar to organisms in undergoing birth and death and in having the capacities of regeneration and reproduction. Although, this metaphysical reading of plant succession is a graphic way of underlining the predictability of the process, it is not essential. The scientific theory of plant succession could survive intact even if the organismic interpretation of it were abandoned.

## *(ii) Problems with the "Classic" Theory*

To accommodate the growing number of exceptions requiring special explanation, the number of variations on the climax concept began to proliferate implausibly, with Tansley at one point seriously suggesting the concept of a "mowing climax." One of the theory's sternest critics was H. Gleason, a younger contemporary of Clements. Gleason took an individualistic view of plant communities, referring to them as "fortuitous juxtapositions" and as "no more than the mass effect of the action or behaviour of individual plants." He found the attribution of overriding agency to climate implausible, and cited existing vegetation, migration, soil, and georgraphy, as equally important factors. In particular he highlighted instances of oscillating succession—between prairie and forest and back again, and between mangrove swamp and open sea, depending in part on whether the shoreline is rising or sinking, so underlining the idea of directional change. Empirical evidence lending support to Gleason includes disparties in the species composition of allegedly similar communities, absence of boundaries between communities, and only fitful evidence of the close-knit dependencies that Clements' theory postulates.

### *(iii) New Explanations*

Later theorists have tended to accept the reality of the phenomenon, but have offered a number of competing explanations of its occurrence. F.F. Egler suggested that existing vegetation often inhibits rather than facilitates new arrivals, and that the order of arrival affects the composition of the resulting plant community. Others propose that most of the phenomena of succession can be explained in terms of the differential colonising abilities of the available organisms, together with their growth patterns and survival capacities. Thus, organisms with high dispersal abilities (*r*-strategists) will tend to predominate in the early stages, while the longer lived, more competitive organisms (*k*-strategists) gradually gain a foothold, and predominate in the later stages. As these explanations indicate, there is increasing evidence to suggest that succession may be governed more by external than by internal factors.

A more radical proposal is that the phenomena of succession may be the consequence, merely, of random or "stochastic" processes; for directional change and repeatable convergence are a statistical property of "Markov chains"—processes in which what happens next depends entirely on what is happening now, and not at all on what has happened previously. The extent that stochastic events may be responsible for ecological processes raises the general question of how far the apparent signs of ecological stability are signs of stability at all; or whether they are just transient patterns in what are basically non-equilibrium systems. Of course, ecological processes need not be exclusively one or the other. Just as stochastic processes can mimic deterministic ones, so can processes that are "chaotic" (but deterministic) mimic stochastic ones—with apparently random developments in fact reflecting extreme sensitivity to initial conditions, which is the mark of a chaotic system. Methodologies for distinguishing between chaotic, stochastic, and deterministic systems are still undeveloped.

## (e) The Diversity-Stability Thesis

### *(i) The Thesis*

The apparent correlation between stability and diversity has been a particular focus of interest, since it first featured as an assumption in the work of Robert MacArthur—to the effect that the measure of diversity could be used a way of estimating the balance or stability of a system. Its appeal rests on a combination of empirical and apriori considerations. Among apriori considerations is the belief that a diversity of interaction relations—for example, feeding options—will make for stability in numbers—for example, of both predator and prey. For predators can turn elsewhere if a particular prey becomes scarce, preventing further losses or both the predator and prey species. Further, it would seem that diversity of characteristics within a given population would present that population with more chances of surviving environmental vicissitudes. Empirical observation meanwhile had suggested:

(a) that species rich systems (i.e, systems with large numbers of species) such as tropical rain forests were relatively stable;

(b) that species poor ecosystems were particularly prone to fluctuations in populations; and

(c) that monocultural systems were relatively unstable.

### *(ii) Problems with the Thesis*

However, both empirical and theoretical considerations have seemed to undermine the thesis. It was noted, for example, that extremely simple systems are also sometimes extremely stable, while extremely diverse systems such as tropical rain forests were proving vulnerable. Correlations appearing to support the thesis could be explained in other ways: in one study where the stability of certain bird populations correlated with their diversity, it was found to correlate also with an east-west and a wet-dry spectrum, and in general with the availability of more suitable habitat (D. Goodman, *The Quarterly Review of Biology* 50, 1975). Moreover, R.M. May showed on theoretical grounds, that diverse systems need not be more stable than simple ones.

### *(iii) Responses to the Problems*

Commenting on May's results, Putman has noted that the type of stability that, according to May's models, is lacking in diverse systems, is constancy rather than resilience, and that some degree of correlation between divesity and resilience might still be affirmed, particularly if the diversity involved is diversity of interaction rather than species richness, for there appears to be no necessary correlation between diversity of interaction and species richness. Further, differences between number and strength of interaction appear to affect inertia, constancy, and resilience in different ways. May himself has gone on to propose that such correlation as is observed in complex natural systems, between diversity and stability is to be explained by reference to a common cause—the stability of the natural environment in which they have arisen and by which they are sustained.

This is consonant with a speculation of Bryan Norton's. Nothing that the concept of stability assumed by the original diversity-stability thesis, in focusing on recovery to some more or less original state, was there fore "static," Norton has used the work of Ramon Margalef to propose a "dynamic" concept. Dynamic stability refers to the autogenic character of a system—that is, the extent to which the later states of the system are a function of, and therefore, predictable in the light of, earlier ones. Persisting species, stable populations, and recovery to some initial state—the marks of static stability—are not required. Dynamic stability, Norton suggests, will be a characteristic not of systems that are internally diverse, but of systems that are themselves formed within a wider context of diversity. This situation will entail intense competitive pressure and therefore result in a system that is highly interconnected internally and likely to be inured to further disturbance. Dynamic stability will tend to correlate with maturity but not necessarily with diversity. Such an analysis will explain how simple systems can be stable, and also how diverse and mature systems can be vulnerable to specifically human pressure, if this should be of a novel kind.

### (f) Discordant Voices

Recent theorising in ecology has seen a resurgence of scepticism about the notion of ecological balance, with disturbance increasingly being regarded as the rule rather than the exception. We have that conceptions of the natural world that emphasise chose and complexity and randomness are gaining ground. Weather systems and randomly generated computer graphics are regarded as more appropriate models for natural processes than clockwork mechanisms, organisms, and homeostatic systems. Empirical considerations at best serve only partly to explain this development. It is in part also a matter of the availability of these new models and in part, too, a matter of residual conceptual difficulties over the concept of balance.

A number of factors may be identified as lying behind the new outlook:

(i) The first is a softening of edges, which is an aspect of natural processes to which certain strands of "deep ecology" have been sensitive. Once the relatively sharp boundaries separating individual organisms are left behind, and once Darwin had disproved the constancy of species, the drawing of both spatial and temporal boundaries around natural processes becomes problematic. Units lager than organisms, such as communities and ecosystems, lack clear criteria of identity. Plant "communities," for example, are found to vary considerably across geographical distances along gradients of elevation and moisture, making the separation of one from the other problematic; and their composition is found to vary considerably depending on the order of arrival of their formative species, making the identification of community types problematic.

(ii) Lack of clear criteria of identity places the notion of "disturbance' in question, because the latter seems to presuppose a clearly identifiable item undergoing the disturbance.

(iii) Lack of clear boundaries places in doubt also the correlative notion of "the surrounding environment," and throws doubt on the validity of the properties that are ascribed to it. There is more than a hint of circularity, for example, in the concept of a "stable environment," which seems partly to gain its sense from the degree of disturbance undergone by that whose environment it is.

(iv) If there are functional, evolutionary, or even historical constraints on the changes that characterise natural processes, there has been difficulty in detecting them. For example, neither the mathematically inspired Lotka-Voltera equations describing predator-prey relations in terms of a "competitive balance," nor the organically inspired succession theory of vegetation, has held up well under empirical study.

(v) Although mere change and variety are not in themselves incompatible with notions of balance, if the change and variety exhibit some constancy or regularity, change in the natural world seems to be inconstant and irregular, and its course to vary greatly, depending on initial conditions. The history of climate change and of vegetational changes as revealed by pollen analysis are both tending to confirm the point.

(vi) Natural events traditionally regarded as "disturbances." such as storms, fires, and floods, are increasingly being seen as creative and integral features of natural processes. Accordingly, there are good grounds for viewing such disturbances as endogenous rather than exogenous features of those processes. Yet their occurrence is erratic, and they must be presumed to pass on this feature to the processes of which they are now seen to form a part.

(vii) There are increasing pressures to "count in" the effects of human activities. A practical reason is that ecology needs to deal with the real world, and separating out human activity is increasingly difficult in view of its global impact. Moreover, the recognition of the erratic character of much natural disturbance makes humanly caused disturbance seem less out of place.

It is possible, however, to exaggerate the significance of these points. Concerning points (a), (b) and (c), it is not clear that the notion of ecological balance requires there to be sharply defined boundaries in nature. If, as some prefer, we speak of "patches" rather than of communities or ecosystems, these may still be regarded as representative *samples* of natural processes—and the regularities and patterns that are found as therefore illustrative of the way the natural world works. Concerning (d), a number of authors have drawn attention to the apparent operation of morphological and physiological constraints in the history of evolution, citing, for example, the fact that different lineages are often found to pursue strikingly similar behavioural and morphological paths. Concerning (e) and (f), neither inconstancy nor disturbance rules out stability. For example, there are known cases where species richness is greater under moderately severe than under slight disturbance; and cases where irregular disturbance keeps competitive species in balance—by preventing one or the other from wholly gaining the upper hand. Concerning (g), it can be argued that humanly caused disturbance is qualitatively different from anything natural, outside of the great mass extinction events. The inference to be drawn from these various considerations may not so much be that nature is revealed as inconstant, irregular and unstable, as that we still do not know how far any of these features obtains, and more radically, that we still lack clear criteria for deciding what counts as stability or instability in the first place.

## 4. THE SOCIAL CONSTRUCTION OF NATURE

Challenges to the idea of ecological balance do not only arise from within ecological theory, but also come from outside, and in particular from perspectives such as the sociology of knowledge that emphasises the extent to which scientific concepts may be culturally determined or shaped. Such a perspective generates apparently oxymoronic talk of the "construction" or "invention" of nature—oxymoronic, because nature is usually conceived as something that is "given." stronger versions of the perceptive will claim that culture determines theory; weaker versions will claim only that cultural factors form an essential ingredient of any comprehensive explanation of why a particular scientific theory or claim is accepted.

Against this framework of thought there would seem to be at least two ways of attempting to preserve some substance to the idea of ecological balance. One is simply to accept the idea of construction and nevertheless insist on the practicality or usefulness of construing nature in this way. But the question is how far the belief that one is actually saying something about nature can survive the admission that this is *merely* a way of construing it. Perhaps, a more promising way, therefore, of preserving a degree of objectivity in the idea of ecological balance, while granting the weaker version of the constructivist point some purchase, is to insist that nature lends itself to being construed in one way rather than another, and cannot in fact be construed in any way we please.

Indeed, the scope for invention in our descriptions of

the natural world is contained from the start in the inherent elasticity of ecological concepts. It has alredy been remarked that ecological processes do not manifest sharp edges and clear boundaries: whether a given event counts as a disturbance depends on where boundaries are drawn. Similar considerations apply to "invasion", "indigenous", "population", "colonise", "alien", and a host of other concepts that appear in ecological theory. A moment's reflection reveals that these terms are parasitic upon the way in which certain boundaries are drawn (as well as being very clearly metaphors with an ideological load); and that a different choice of boundary will entail a different reading of the event. Other sources of elasticity steam from the nature of the tools and instruments used to process ecological data. The point is nicely illustrated in a recent paper by P. Alpert, who describes how the boulders that formed the actual habitat of the mosses that he studied came inevitably to be modelled as spheres during the process of analysis. He shows also how much the results that he produced were a function of the choice of hypotheses that he decide to test and those that he chose to leave alone.

What survives such revelation, however, is the fact that the boulders lent themselves more to analysis in terms of spheres than—say—rectangles, and that the results could at least be justified in the light of the hypotheses chosen. To the extent that nature is "socially constructed," there would appear at least to be a certain construction manual. Hence, the "constructivist turn" need not take the ecologist entirely by surprise; many ecologists will take the view that such elasticity in their findings is endemic to their activity. They might well agree that, given the slack between theory and data, it is not unreasonable to suppose that cultural and psychological factors will help to explain why one theory rather than another is chosen, and why a third, perhaps, is not even considered.

## 5. HUMANS AND ECOLOGICAL BALANCE

Judgements about the impact of human activity upon ecological balance go a long way to shaping not only current perceptions of our environmental predicament but also the image that human beings have of themselves and of their lives. It raises, for example, the question whether humans are, or are not, to be considered "part" of nature, and how far they can consider themselves to be "at home" in the word: it matters to us to know what kind of world we are thrown into. In fact, humans pose a problem for the thesis of ecological balance inasmuch as the part they increasingly play in relation to the rest of the biosphere is in many ways unprecedented. The single fact, for example, that being only one species they nevertheless appropriate about 40% of the terrestrial biomass, is prima facie difficult to assimilate with in that thesis.

There are three responses to this phenomenon that would save the thesis. One response is to treat humans as separate from the rest of nature, and therefore as falling outside the scope of the thesis. But this is a difficult position to maintain in the light of Darwinian theory. At what point exactly in the history of evolution did the separation occur? Another response is to postulate at some future date some answering violence or correction that may or may not spell the end of the human race, but would restore the biosphere to some kind of balanced state. A third response is to place humans firmly within the natural world but construe their apparently uncontrolled growth as a form of disease. Earlier writers had sometimes likened the human race to a plague or pestilence with respect to their effect upon other species. Against the background of a belief in nature as a system inbalance, a more favoured analogy has been to see humans as a disease of the system—specifically, cancer. Carleton S. Coon, for example, in his *History of Man* refers to mankind as "that terrestrial skin-cancer"; and a contributor to the *Gaia Atlas of Planet Management* refers to humanity as a "super malignancy." Coon is equally clear about the mechanism by which humans upset this natural balance: it is through the practice of artificial selection and the "thwarting" of the processes of natural selection. Other analysts have not been slow to add to the list of symptoms. Attention has focused most notably on the way in which human activity ruptures the existing food-web relations, and thereby also the energy and nutrient cycles. It has focused also on the loss of biodiversity.

Where humans are not cast as themselves the disease, they are cast as being responsible for disease. Thus, Leopold writes of the wounds that only ecologists can detect and calls upon them to practice the science of land health. He saw the conservation schemes that had so far been attempted as merely "local alleviations of biotic pain." This last image at least presents humans in a slightly less negative light. For while they are accused of being responsible for the patient's illness, they are also seen as in principle having the opportunity to effect a cure—in their capacity as "land doctors."

## 6. ECOLOGICAL BALANCE AND NATURE CONSERVATION

The idea of ecological balance has been of major importance for nature conservation. This is because, to be true to its name, if nature conservation is about maintaining stability of any kind, it must be about sustaining *nature's* stability, or trajectory, If, therefore, the idea of nature's stability or trajectory is placed in doubt, ideas about nature conservation that presuppose such a stability or trajectory must be in question, as must the very idea of nature conservation itself.

**Table 2.2**

**Conservation Objectives**

| *Initial state of affairs* | *Appropriate response* | *Mode of response* |
|---|---|---|
| Intact | Conservation | (i) Keep things the same |
| | | (ii) Keep things going |
| | | (iii) Let things be |
| Damaged | Restoration | (i) Undo the damage |
| | | (ii) Allow "natural" regeneration |
| | | (iii) Restore to "natural" state |
| Destroyed | Compensation | (i) Create the same thing elsewhere |
| | | (ii) Create an equivalent elsewhere |
| | | (iii) Create a substitute |

Conservation objectives that in one way or another assume some pre-existing state of ecological balance can be tabulated (with admitted simplifications) in the way indicated by Table 2.2, depending on whether the site is viewed as relatively intact, "damaged," or virtually destroyed.

Such as schema is not free of internal difficulties Consider the third mode or response to damage—restoration to some natural state. This might suggest restoration to how things naturally were—which may be impossible to effect in a changing situation; restoration to how things naturally would have been—which may be impossible to predict; or restoration to how they might naturally have been—which makes it unclear what exactly is ruled out.

There may also be some immanent tension between two of the "guiding stars" of conservation thinking that is broadly governed by the assumption of balance, namely, ecological health and ecological integrity—and this despite their both having the imprimature of Aldo Leopold. For whereas health is a functional concept, integrity contains the idea of being true to origins. Like sustainability, both concepts are still undergoing critical development, but picking up a suggestion of J. Baird Callicott's, we might usefully think of health as being more appropriately predicated of ecosystems, and integrity of communities. Whereas communities are partly defined by reference to the populations and species of which they are composed, ecosystems are functioning entities not requiring the persistence of any particular population or species for the functional role to be maintained. But herein lies the source of tension that at present is manifest in debates about the proper conservation approach to so-called "exotic" or "alien" species. A number of extermination schemes are in operation designed to protect indigenous species against "invasion"—rats and ruddy duck being among the target species. Although, this is usually billed as a conflict between conservationists and animal welfare groups, it is at least as much a conflict within conservation thinking itself. For while rats and ruddy duck may well be a threat to integrity, they are not obviously a threat to health, especially if they perform a similar role to that at of the species they replace.

It seems likely that any relaxation from the governing model of pre-existing balance, although it will not remove sources of conflict, will permit more creative responses to environmental problems. Creative conservation, as it is sometimes called, looks to "future nature" and "new nature" and tends to be skeptical of the human/nonhuman divide. It questions the system view of nature and associated notions of stability and health, but on the other hand will also tend to being conflict with approaches that stress the importance of the indigenous and the rare. It will be in tune with the notion of nature as an open-ended historical process and be ready to come to terms with a nature that is unpredictable and wild. It will view nature conservation as less a matter of protecting nature's balance from humanity's destructive power, and more a matter of preserving the destructive wildness of nature from humanly imposed stability and monotony. But if "anything goes" in nature conservation, on the grounds that "anything goes" in nature, skeptics will worry lest it legitimates practices that, as was famously said of politics, merely prosecute the old war against nature, but by other, more diplomatic, means. Accordingly, many conservationists will still large for the existence of real functional, and historical constraints as being characteristic of natural processes, and therefore factors that nature conservationists must heed. Their fear is that creative conservation, as described, will inevitably lead to homogenisation and loss of biodiversity within and among regions.

There are recurring patterns that characterise the natural world at the level of individual organisms, both in the conformity to type which grounds the notion of species, and in the pattern of individual development. Because this recurrence is partly constitutive of life, and because life does not and probably cannot occur in solution, it seems reasonable to suppose that collections or groupings that incorporate living things will also exhibit recurring forms. Thus, life and collections of living things both presuppose and exhibit a degree of relative stability. And just as individual organisms possess various degrees and kinds of recovery capacity, from the healing of wounds to the regeneration of whole limbs, so it also seems reasonable to suppose that collections or groupings of living things will constitutionally possess similar powers of recovery. In short, it may be said that life itself is "a balancing act." But from the fact that, if there is life, it will manifest balance of a certain kind, it does not follow that nature manifests a kind of balance that will ensure the continuation of life in all but the most extreme of circumstances.

The abandonding of providential ecology appears to mean that ecological balance can no longer function as an indispensable explanatory assumption, but itself stands in need of explanation. The issue is whether it can be shown to be an inevitable concomitant of those factors now perceived to determine natural processes, or whether it is nothing but a contingent epiphenomenon of those processes.

The elegant swan looks different viewed from above and below the water line. To a considerable extent current disagreements over the balance of nature may reflect where people choose to stand and where to direct their attention. To a certain extent they may also reflect the room for cultural construction or invention allowed for in the application of ecological concepts. But a deeper question is whether and to what extent physical and biological processes may be reducible one to the other, or whether one must settle for an irreducible plurality in scientific explanations. On a pluralist view, the issue of ecological balance can be seen as an empirical question to be provisionally decided on in the light or prevailing ecological theory. On a reductionist view, the notion of balance might or might not be confirmed. The observed regularities are there; but where they manifest a real ecological balance depends on how far they are underwritten by the fundamental laws of nature (whatever they may be). For example, one can speculatively imagine a mediating theoretical role here for the concept of the gene. To the extent that the distribution of genes turned out to be some function of the chemical affinities of their material substrate, and to the extent that species distribution turned out to be some function of the distribution of genes, then physico-chemical processes could conceivably turn out to determine ecological regularities. But whether reductionism is true is a hard question to resolve, and for as long as questions of this kind cannot be decisively resolved, whether and in what sense ecological balance is a feature of the natural world is likely to remain unresolved.

## REFERENCES

Begon, M., Harper, J.L., & Towsend, C.R. (Eds.), (1996), *Ecology* (3rd ed.). Oxford: Blackwell.

Botkin, D. (1990), *Discordant Harmonies.* New York: Oxford University Press.

Brennan, A. (1988), *Thinking about Nature*, London: Routledge.

Callicott, J.B. (1995), The Value of Ecosystem Health, *Environmental Values*, 4, 345-361.

Costanza, R., Norton, B., & Haskell, B. (eds.), (1992), *Ecosystem Health: New Goals for Environmental Management.* Washington, D.C., Island Press.

Feidler, P.L., and Jain, S.K. (Eds.), (1992), *Conservation Biology: The Theory and Practice of Nature Conservation, Preservation and Management*, London: Chapman and Hall.

Golley, F.B. (1993), *A History of the Ecosystem Concept in Ecology*, London, Yale University Press.

Johnson, K.H. *et al.* (1996)., Biodiversity and the Productivity and Stability of Ecosystems, *Trends in Ecology and Evolution* 11, 372-377.

May, R.M. (1986), The Search for Patterns in the Balance of Nature: Advances and Retreats, *Ecology* 67, 1115-1126.

Norton, B.G. (1987), *Why Preserve Natural Variety?* Princeton: Princeton University Press.

Noss, R. (1996), Sustaining Ecological Integrity, *Global Bioethics* 11.

Pimm, S.L. (1991), *The Balance of Nature,* Chicago: University of Chicago Press.

Putman, R.J. (1994), *Community Ecology.* London: Chapman & Hall.

Soper, K. (1995), *What is Nature?* Oxford: Blackwell.

Westra, L & Lemons. J. (Eds.), (1995) *Perspectives on Ecological Integrity,* Dordrecht: Kluwer.

Worster, D. (1994), *Nature's Economy* (2nd ed.)., Cambridge: Cambridge University Press.

3

# MORAL DILEMMA OF ECOLOGICAL CRISIS

No doubt all living beings have their particular homes in Nature to which they are tied by various strands of environmental relationships of interaction, co-operation and even disinterested neutrality which condition and influence their development, reproduction and survival. Thus on the integrity of this complex web of biological processes depends the survival of all living beings including man. Here, considering the high position man holds, he has a terrible responsibility in the system, as his abuse, overuse, carelessness, or waste can bring the collapse of the entire system, which does include man himself. Responsible ecological security thus involves concurrent and integrated care of all the components, the conservation, responsible and wise management of the life-support system of land, water, flora, fauna and the atmosphere.

Man's greedy egoism and his obsession for more luxury, more comfort and more profit by means of science and technology has made him selfishly exploit the environment and upset the ecological equilibrium and relationship so that the earth which ought to be man's home has become almost like a gas chamber. As one ecologist said: "Civilised man has marched across the face of the earth and has left a desert in his footprints." The ecological devastation was such that it now hits back at man in such a manner that not even the most sophisticated weapons he has made is able to counter its onslaught and it now threatens the life-support system essential for his survival. Human existence is sacrificed to the advantage of blind technological progress and unbridled appetites for luxury, profit and power. "Every time we produce a Cadillac we do it at the cost of decreasing the number of human lives in the future."[1]

Man has become an endangered species along with one thousand other species. This is the frightening ecological crisis, perhaps *the* crisis of contemporary world that has assumed alarming magnitude and complexity in the most recent decades. The crisis is mounting feverishly and is tolling the death-knell of mankind on this planet. It poses a great threat than even nuclear warfare or unrestrained militarism, for, while the latter is still in theory within the conscious control of human decision-makers the ecological threat, poisoning and depriving air, water and soil, is invisible and go largely unnoticed, and requires for its prevention actions that will have to go against deep-rooted social value, lifestyles and economic systems and demands global efforts unknown to the world history.[2] Ecologists have raised the alarm and issued serious warnings regarding the ecological crisis:[3] If man is not becoming conscious of the way this crisis is threatening his very existence, it will overwhelm him like a deluge laminating all traces of his existence from our planet. Advocating patience here is an invitation to be spectator at our own destruction.

In this essay, we wish to deal with the ecological crisis and the consequent problems or more exactly and modestly some of the problems of the ecological crisis. These problems are seen in terms of a dilemma and of the possible responses to solve it or to escape from it. The dilemma in question is: the more we depend on technology, the more we are thrown into the environmental crisis—the more we produce, the more factories we build, the more we are subject to new dangers of ecological catastrophe. Winning the battle with Nature man finds himself on the losing side.[4] The scientific achievements that released mankind from the claws of poverty, hunger and disease have now become the most lethal weapons threatening the very human survival.

The situation of the poor Third World countries is more desperate. Faced with extreme poverty, hunger and overpopulation, these countries are caught up in the dilemma: on the one hand, the necessity of making use of "`extreme" technologies (which are highly eco-destructive) for their economic growth to solve their poverty problem. Curtailing the technological benefits and substitution them with less harmful ones would bring massive economic burden on them, as much of the productive economy of the Third World is linked precisely with this 'high' technologies. And on the other hand, the introduction and spread of more 'extreme' or 'effective' technologies and industrial development in areas like agriculture, forestry and fishing in these countries lead to exploitation, and destruction of scarce natural resources, seriously undermine the survival of the poor and marginal communities, who depend on these resources for their livelihood, cause large scale unemployment, and uprooting of people and their cultures, besides atmospheric changes.[5] These poor nations also fear reprisals of the powerful industrial nations, who, by cutting off the imports of their goods could easily destroy their already weak economy. Added to this there is always the danger that through new inventions or discoveries, the resources that they possess and are at one time in great demand will no longer be needed.[6]

The dilemma is deeply rooted in a number of complex problems like poverty, population and food in the Third World countries. As R.A. McCormick says, our eco-catastrophe is constituted of too much people, too much use of resources and too much pollution.[7] Of the present 5.3 billion people on earth, about a billion live in abject poverty. We do not know the limits of our biosphere. Can the earth meet even the modest aspiration of this without irreparable damage to its life-supporting systems? Any solution that tries to tackle ecological crisis or population growth or the food problem singly is bound to fail now more than ever before, as these problems with its concomitant phenomena of urbanisation and economic growth-oriented industrialisation and development are inextricably linked to each other. We are facing one problem, not three different problems.

Hence the question, for example: Shall a pesticide that might save crops for hungry mouths not be used because of what it will do to the people for forty years from now? Shall a river be dammed for its ability to irrigate and power a hydro-electric plant? Is present generation or mankind obliged to be concerned about the problem of future generations, the amount of required energy, food and natural resources necessary for their well being? The main question is whether we have the right to put an additional fearful load on generation yet unborn. This is the moral dilemma with which we are caught up and our choices are not easy. With some choices like nuclear energy the whole planet may be at stake. Are we to sacrifice then development and growth for the sake of environment? The poor countries cannot afford such a step since industrialisation is one of the basic needs of these countries. The question is how best can we adapt our environment through technology to serve our real needs. It is a question of moving through a delicate balance between the use of technology and protection of environment. But is appears that the use of 'high' technologies will continue in accordance with the greedy exploitation and massive destruction of nature which will ultimately drive us into the ecological catastrophe wherein our very survival will be at stake. "Our dilemma is an unfortunate outgrowth of Judeo-Christian heritage which has produced a blind science and technology and a berserk econo-centric culture."[8] And this calls for theological and ethical responses.

## ETHICAL RESPONSES TO THE DILEMMA

Based on an authentic Biblical and adequate theology of ecology.[9] we need to formulate not merely man's ethical obligation towards Nature and all living beings, but articulate a socio-economically just ecological ethics—an ethics based on a just and responsible stewardship of the earth. This is important, because ecological crisis is not merlely a problem of environmental pollution but a problem of socio-economic injustice.

### A Socio-Economically Just Ecological Ethics

The rutheless exploitation of nature is always linked with exploitation injustice and of man by man. And more devastating than an enviromental pollution is the pollution of injustice and the consequent pollution of poverty and misery in world today. Already we have a situation where 500 million people in the world exist in a state of chronic malnutrition and 8000 people starve to death each day. An adequate, effective, and liberative ecological ethics should include the question of justice due to the deprived poor and responsibility for a healthy and habitable environment for *all.*[10]

The affluent countries are exploiting and depleting the scarce resources of the earth and polluting the environmental an excessively high rate than the poor countries. Exploiting underdeveloped nations for their resources has always been a characteristic of the developed nations. The rich and industrialised nations who form only 1/3 of humanity wastefully devour 83% of all material resources leaving only 17% to 2/3s of humanity who live in the Third World on the edge of poverty and many deep in its cruel grasp.[11] Every second the world uses 30,000 gallons of petroleum. U.S.A. uses about 10,000.[12] The U.S.A., the Soviety Union and the European countries ranked number one two, and three respectively for the contribution of dangerous, heart-trapping 'green house' gases added to the atmosphere.[13] This situation will continue to the detriment of justice to the poor, because the power to decide on policies and priorities in industry, agriculture, and technology, remains concentrated in the hands of these nations. And in the poor nations, it is the ruling elite technocrats, industrialists and greedy landlords who dictate the scandalous priorities in these sectors, depriving the basic needs for the landless, labourers and the little peasants.[14] There is a marked deference in the ecological concerns of the affluent, polluting and exploiting countries and persons and the victims of such exploitation.[15] While the ecological concern of the rich are to keep their level of luxury and affluence, threat of the poor is one of survival and basic needs.

Now the ecological imperatives advocated by the ecologists of the West or by the rich people are often biased in their favour and do not mention the question of justice. Their debarring of any intervention on environment or counselling a growth-freeze hits the poorest of the poor, who struggle for basic needs and whose only hope for betterment lies in a growth situation. Growth-freeze will only institutionalise poverty for many. If there has to be a choice between justice or conservation, then justice must have the priority. Social ecology is more important than physical ecology. Also some anti-pollution measures, creation of some pockets of greenery, wildlife preserve, etc., are only ecological palliative and movements including the nature-ethic and reverence for life are fundamentally bourgeois in character.[16] They do not touch the heart of the ecological question: the issue of domination and the question of justice and responsibility for the care for one's fellow men whereby caring for their concerns and the world they live in. "To talk about the environment in isolation from the massive world problem of injustice, oppression and poverty would be profoundly irresponsible."[17] Instead of saying: "you shall not create an imbalance in the ecosystem of nature," it would be better to say: "you shall not do this irresponsibly."

This brings us to the next issue—the concept of private property. If we want to know the socially responsible way to treat the earth and its products we must take into account certain fundamental ethical, obligations that go with ownership and use. Man does not enjoy absolute ownership of the natural resources, so that he can misuse, destroy, neglect or want only waste. Ecology of human relationships always suffers from such abuse. Further, equal right of ownership leads to the question of the right use of resources. Resources have limits. That brings the problem of distribution of resources. Many people think ownership ought to involve unlimited right of disposition, and that is what they mean when they speak of the 'sanctity' of private property.[18] The Bible assumes not the 'sanctity' but the social responsibility of private property. Applying this to the ethical disposition of earth's resources, their use and conservation, we should remember that those who currently control them people or nations, do not have unlimited right to their use. All men have a right to the means of subsistence, a moral axiom which qualifies the ownership of resources necessary for life. Goods of the earth are destined for all mankind. "What justification is there for a group to have a high level of luxury at the price of hunger for many others? We have to rethink the ownership laws, if all mankind's right to life is guaranteed."[19] And as John A. Ryan says: No one has the right to luxuries so long as anyone lacks necessities."[20] The greed that keeps so many members in the southern hemisphere is of a piece with the greed that pollutes environment.[21] Mahatma Gandhi once said that there is enough for every man's need on this earth, but not enough for every man's greed. Real explosion is not population but the consumption explosion of developed nations. And as Pope Paul VI said: "it is inadmissible that those who have control of wealth and resources of mankind should try to resolve the problem of hunger by forbidding the poor to be born." Population decrease comes most effectively through economic development and there is something ecologically inhuman and unbalanced for a government to enter forcefully the sanctuary of the family, when economic justice and more organic development are left unattended.[22] Certainly any lasting solution to the problem of world hunger seem to depend on improving local self-sufficiencies, but the stubbornness with which the wealthy Northern nations cling to the current world economic system, in which much of their affluence depends on arrangements almost sure to keep the Southern nations poor, tilts the whole problem.[23]

Human bondage is partly a result of this greedy expropriation of the earth's resources by a few affluent people and countries. As T.S. Derr says there is no true fullness of human liberation without also a liberation of the ecology of the earth's resources for the benefit of all men and for the full humanisation of the cosmos. He does not mean that there should be no intervention on

Nature, but thoughtful and judicious, based not for its selfish exploitation but in order to husband it for its own well-being and for the qualitative humanisation of man.[24] There is need for man to consider himself not only as a consumer, but also as a co-creator of the natural resources. What is required is a mastery of man's greed to exploit rather than the mastery of the earth.

To plunder the earth wantonly, to waste its resources in feeding a consumerist culture, to disfigure its face and little it with plastic abominations, to pollute its water and air, are not only irrational and criminal acts, but acts of unbelief, refusal of commitment to God—a type of behaviour quite unfair to future generations, to the earth unfair to God, creator of the earth. Far worse injustice and unbelief is to grab and fence off the best part of the earth and its wealth for the exclusive benefit of a few individuals or groups or corporations or ethic circles or armed band.[25]

Even if we cannot sacrifice the living for the unborn,we cannot defend the life style of the currently affluent, where it robs future generations of basic necessities.[26]

The environment is a unique patrimony belonging to all mankind and is to be sustained for the well-being of all. God destined the earth with all that it contains for the use of all men and nations, in such a way that created things in fair share should accrue to all men under the guidance of justice tempered by charity.[27]

The rich nations should share the resources across the face of the earth, not in charity, but in true justice. The environmental crisis calls for a human control over nature, wise management and distribution which is more than man's capacity to discover and exploit nature.

An earth which is so managed as to be ecologically sustainable is of more value to humanity than one which is ecologically unsustainable. An ecologically sustainable earth is a necessary requirement for distributive justice.[28]

Intrinsic to distributive justice is the issue of natural resources and the wise and responsible use of it.

A socio-economically just ecological ethics demand making the world a decent habitat for our neighbour, i.e., to keep his air, water, food from chemical poisons, give him a place free of chemical poison and give him a place to live in the sun which is not hidden by smog. It must make clear statements on the 'ecological sins,' the greed, the way-wardness, the wasteful habits based on selfish values such as seeking one's pleasure luxury and leisure at the cost of the needs of the poor and downtroddden. It must condemn the unjust profit oriented, consumeristic economic structure of the First World that keeps the Third, World continually in poverty.

*Eco-ethical Perspectives in Science, Technology and Progress*

Science and technology which liberated mankind from the tyranny of Nature have now become the most lethal weapons threatening our very existence. They have lost their integration and direction, and have become a great power which rules man today. This is because in science and technology, there is an intrinsic ambivalence which comes due to the fact that our actions are done in space and time and we cannot control the negative effects of our activity in the world.[29] This ambivalence of either enhancing or endangering our freedom is alway present in the use of technology. Every scientific and technical gain is to be bought with some loss elsewhere. Today's large scale disasters and environmental pollution is an intrinsic feature and clear manifestation of this ambivalence. Science and technology give benefits but they are also experienced as a threatening power. They possess an intrinsic 'ethical' power in themselves. They are no more neutral and value-free. "Technical instruments do not only have the innocence of a means, leaving the ethical decision to our free choice, they possess their own ethical qualification"[30] And this calls for a new and heavy ethical notion of responsibility, not simply to choose for the good but, to save the good itself from the scientific and technological determinism and oppression.[31]

Ecological crisis has questioned the scientific and technological optimism and their ethical neutrality and calls for a serious revision of their presuppositions. We can no longer accept the present system of science and technology based on the old mechanistic world-view that is anti-ecological and the 'interventionist mentality' that shape our moral perception—the bias of a technologised culture to believe that interventionist manipulations are genuine (i.e., human) solutions to the problem.[32] The crisis calls for a new orientation of science and technology towards an integrated holistic vision that sees the real and genuine human needs, and puts them at the service of promoting the quality of life and happiness of man. We have to preserve some 'free zones' in our lives, where technology has not the unique and certainly not the last word. We believe this is possible, if only we do not invest our technical means with the status of goals; then we can control their negative forces.[33]

This is often neglected by the experts in science and technology which is spiritually impoverished and divorced from purpose and values. What matters is therefore not merely the ethical stance of the human beings who use the instruments and choose the goal. It also depends on the spirit of both scientist and technologist who must restore the religious and ethical values and goals to science and technology. The purpose and values of science and technology must be theologically and eco-ethically reoriented, i.e., not as to rule over Nature, but as to perform their duty, which is to guide human interactions with rest of Nature, thus preserving and protecting its integrity.

Scientists should resist and not conform to governmental policies which may be politically and economically expedient but are tantamount to ecological murder. "Scientists must find ways to move the restrains of secrecy and insist on public discussion of the possible consequence of large scale experiments before they are undertaken."[34]

But, unfortunately, today 40% of world's qualified scientists work in industries that cause major environmental damage and senseless arms race. Right now over 400,000 engineers and scientists arc engaged in ecologically insane and senseless military research.[35] Do we really wish the world to continue at all? This last question is really ethical and it really matters. Care for the future of mankind is indeed a distinctive feature of the person as a moral agent. Our responsibility is to counterbalance that threat by creative effort to discover new dimensions of Nature (cosmological and ecological), the meaning of solidarity and the ground of our existence. It is our responsibility, mainly on our own behalf, to reduce the negative aspect of technology's specific ambivalence: we should not let ourselves be programmed by technology. The ethical aspect of science and technology asks for this added responsibility of much self-denial, and insists on priorities based on responsible justification and proportionate reason.[36] We have to make our choices so that we also include in our will the future integrity of the human beings. "Man, as co-creator under God, has a right to manipulate the processes of Nature wherever he can judge with moral certainty that his action is beneficial for the present generation without jeopardizing the future one."[37]

And because of the largely known and unknown

consequences of interventions the deontological element must have priority in the evaluation of the possible consequences so that we do not fall prey to strict consequentalism. If there is anything meaningful in Christian ethics, it has to invite modern science and technology to discipline itself to serve all humanity, not personal success, not the ubiquitous power-play in the world.[38]

When 500 million people are suffering from malnutrition and 8000 people are starving to death each day, we cannot renounce science and technology altogether. What we need today is not "environment or development" but "environment with development." We have to strike, a balance between environmental protection and development, as both are absolute necessities in the present context. As Moltmann says, quoting the Chinese proverb: "He who rides on the tiger can no longer get off it, so it is better to tame this tiger and with it to reach the future."[39] Science and technology must be directed with a wisdom and control for the benefit of all mankind. A more humanistic science with an ecological wisdom that co-operates with nature rather than trying to master nature is required. This means choosing 'appropriate technology' along ecological principles that contributes to the harmonisation of society and nature. (e.g. the 'soft' technology for energy needs with renewable or inexhaustible resources, like sun, wind, etc.)[40] Also the 'intermediate' or small scale industries after the example of Mahatma Gandhi are beneficial in developing countries where there is tremendous labour surplus.

'Hard' technologies destroy creativity, causes unemployment and increase environmental pollution. The technology that we choose "will determine much of the future shape and quality of our life—not just in terms of pollution, but in the very make up and organisation of our communities and our political structures."[41] In the conceptual focus of 'basic need' and 'appropriate technology,' we can find the goals for the future of both rich and poor countries and that makes good ecological sense as well. Somewhere between the bullock-cart and the nuclear power plant there would appear to be a more sane path into the future for all of humankind.

*Human Progress:* An eco-ethical perspective would demand a revision in our concept of human progress. Progress should not be viewed purely from a quantitative and materialist point. Increasing success of science and technology has led to an ideology of developmentalism. This questionable ideology equates human progress with technological progress. We should consider progress in terms of the quality of life, oneness, harmony and wholeness—"development more in being-becoming than in having-consuming,"[42] We have to learn that true human progress does not begin with exploitation of goods of profits but with people, their education respect for their dignity, equality, justice and discipline. Man who exploits nature for selfish ends is destroying the best in himself—the sense of responsibility.[43] True human progress does not belittle technological development—but then it is technological development which is at the service of the whole creation. To put it in theological terms, science and technological development should be a participation in God's total creative activity, which can truly be so, only if it is born out of love—love for the world, the whole creation, which is nothing but God's own self-reflection. Such a concept of progress calls for a 'bottom up' approach to development were the scarce money and resources are used for the basic needs of the poor rather than squandering them on luxuries that our planet may literally not be able to afford. The materialistic or profit-oriented, quantitative model of development which is linked to the exploitative use of science and technology is causing depletion of non-renewable resources and pollution of the environment, dehumanising it instead of humanising.

## ECOLOGICAL POLITICS

Ecological crisis has become a serious political issue today. It cannot be solved by good intentions nor by heightened awareness nor technological innovations alone. All these cannot possibly reverse the present logic of environment unless the overriding political, social and economic framework which has actually generated, it is radically transformed."[44] As long as there is a dollar or yen or some other advantage gained by ignoring long term consequences, there will be economic powers and political powers (who are often manipulated by economic powers) who will engage in colossal damage and massive exploitation of the environment. Nor can the problem be solved segmentally as it is global. The magnitude of the problem demands a holistic approach that calls for the mobilisation of worldwide solidarity and political will for globally responsible political actions. As the cause of wastefulness and reckless destruction of our ecosphere is in large measure in western thought, economy, politics and established habits, much depends on the people and political leaders of Western nations who are in a position to make policies and commitments for a just and pollution-free new world.[45] The people of rich nations should be made aware that if environment is protected in their countries they are also going to benefit from it. If forests and other precious resources are destroyed in one country, the effect of that on environment will not remain confined within the boundaries of the country alone. Pollution does not recognise national boundaries. This should make the rich nations reduce the debt burden on the people of developing countries. Frightening ecological threat could become another motive for removal of gross inequalities.[46]

It is relatively easy to speak of guidelines. Are wealthy nations prepared for radical economic changes which drastically affect their economic structure in order to resolve the gross international inequalities?

International politics today pivots on economic greed. A capitalist, consumerists and profit-oriented economy, industry and social organisation promoted by the affluent countries and the powerful elite of the Third World countries determine the quality of our environment. The politics of every other environmental issue is the same: a tiny majority versus a huge majority. And as political decision-making regarding our biosphere is in the hands of few economic powers, ecological priorities will be distorted and will be biased in their favour. And as Bernard Harving says the few multinationals with their powers and with their experts in all fields influence public opinion in ecological matters. "By buying television time, by lobbying, bribery, blackmail and especially by the argument that jobs of thousands of people depend on them, they manipulate legislators and public opinion."[47] Wherever environment is looked upon as the toy or the pawn of political economic powers and where a democratic, environmental policy has not yet evolved the problem will continue to aggravates.

The politics of ecology must start from the premise that the environmental crisis is increasingly the product of a structure of economic and political power. Radical political decisions with regard to private property, science and technology, industry, agriculture, forestry, fishery, transport, etc., are to be taken both on national and international levels. There is need for technological development, if we are to be self-reliant, but not at the expense of massive exploitation of natural resources. Compromises in terms of intermediary programmes and structures may be inevitable. But such compromises taken by governments must be honest and realistic.[48]

For all these, we need democratic environmental policies guided by global ecological awareness and commitments that integrates ecological values. We must raise the issue of pollution, exploitation of nature, wasteful styles of consumption to a political status. And there must be participation of people in the decisions concerning the use of technology and science. Our biosphere and the decisions concerning this life-supporting system is too important to be simply left to the whimsical decisions of political and economic powers and of the experts. Caring for our biosphere is not an optional activity of a few. It is a collective task. The public has to be conscientised about ecological facts and their implications. There should be increasing interaction between ecology and public affairs. The public morality ought to evaluate and determine at what cost we enjoy freedom from pesticides, the convenience of automobiles, nuclear power, etc. Every human being must asset his ownership and his rights over land and air and water which are public property. No industry or company has the right to poison them.[49]

Only through activating people at grass-root level and their participation in environmental issues, the primacy of social justice in political, economic relationships can be guaranteed. What we require are critical and forceful initiatives that challenge the logic of domination operative in our society, that generates environmental problems and bars the poor from having access to the natural resources for their basic needs. Also we need intelligent moves to subordinate technological developments to the goals of justice, to the production of essential goods for the poor and the unprivileged.[50] At the same time, we require people's solidarity and concerted political actions, programmes and waves of ecological movements like the 'Chipko' movement at micro and macro level to overwhelm and force Governments and legislators to deal responsibly with ecological problems by adopting stricter measures with regard to the location policy of polluting industries, disposal of hazardous wastes, etc. A just political ecology, critical of the profit-oriented economy and direction-less science and technology, should be the goal of all these movements. The redemption of people involves the redemption of the world they live in. The bonds that confine people all over the world are economic, political and technical as well as spiritual. The struggle for liberation is a struggle for economic, political, ecological and spiritual liberation. It is vital for the Churches to be involved, boots an all, in all these tasks and to question seriously their commitment to the technically dominated society.[51]

## ECO-SPIRITUALITY

Our responses to the ecological crisis must encompass a revolution not only in political institutions and economic relations but also in our life style and sprituality. If we are to survive this crisis, we have to live and promote an integral eco-spirituality, forming in us an ecological psyche—a spirituality that is creation-centred with an ecological conscience that questions our ego-centric lifestyle and criticises our ideas of 'homo-economicus' such as free enterprise, gluttonous consumption and destruction of our biosphere. The crisis calls for a spirituality wherein the divine mystery is experienced in our deep communion with nature—a creation-centred spirituality that elicits our wonder and awe, encourages new humility within the total earthly community, and calls for an urgent ethical commitment as co-creators and responsible stewards of the preservation and enhancement of Nature. Such a spirituality is in line with the kenotic spirituality of detachment and renunciation characteristic of Indian tradition with an ecological *ahimsa* (non-violence and respect for life in all its forms) *karuna* and *maitri* (compassion, care and kindness for all beings) that tames our violence and exercises our demonic destructiveness to nature.

Such a spirituality is required today to insist in uncompromising ways for a profoundly personal level ecological conversion, from an exploitative consumeristic attitude to a 'reverend-receptive' attitude coupled with a simple lifestyle. This would bring "a renewed relationship with God-given nature—but always with and through renewed relationships with humankind."[52] We need to bring together in our spirituality the profound and the mystical conviction that: "When we strew the landscape with garbage, we taint another portion of our soul."[53] We participate in our society's obscene wasteful ways more than we care to admit. There is a mounting consensus among the ecologists that the world has fifty more years of existence, if we do not introduce radical changes in our consumeristic lifestyle. We need to live simply the Gospel lifestyle of Jesus typifying the Beatitude and social responsibility, so that others might simply live. "It is an obligation of Christian faith and justice that everyone adopts this lifestyle on voluntary simplicity, if poor are to have a just share of the limited resources of the earth."[54] As we emphasised, ecological responsibility is an act of justice to our fellowmen and an act of praise and worship of God.

## CONCLUSION

The solution to the ecological crisis is an immense and gigantic task. There is need for a worldwide solidarity, political will from political and economic powers and a radical conversion of egocentric life, thinking, behaving, ideology, economy and lifestyle. The above ethical and spiritual responses can help us bring back that harmony and wholeness between man and man, man and Nature, and both with God; that is the liberation and salvation of the whole cosmos.

## REFERENCES


1. Nicholas Georgescu-Roegen, "The Entropy Law and the Economic Problem," in Herman E. Daly (ed.), *Toward A Steady-State Economy* (San Fransisco, 1973), p. 46.
2. Robert Dish (ed.), *The Ecological Conscience: Values for Survival* (New Jersey, 1970), pp. xiii, xiv.
3. Thomas A. Sancton, "Planet of the Year, What On Earth are We Doing?," *Time International,* Vol. 133 (January 2, 1989): 16-19; John Langone, "A Stinking Mess," *Time International,* Vol. 133, no. 1 (Jan. 2, 1989): 32-33; John Carmody, *Ecology and Religion: Toward a New Christian Theology of Nature* (New York, 1983), pp. 13-27; Lewis Mumford, "Closing Statement," in *Ecological Conscience, op. cit.,* 100-106; Jurgen Moltmann, *Man: Christian Anthropology in the Conflicts of the Present,* Philadelphia, 1974, pp. 20-25.
4. E.F. Schumacher, *Small is Beautiful* (New York, 1975), p. 14.
5. Felix Wilfred, "Nature and Human Survival," *Jeevadhara* 18 (1988): 58.
6. Barbara Ward, "A New Creation? Reflection on the Environmental Issues," *Pontifical Commission, Justice and Peace,* (Vatican City, 1973): 47-8.
7. Richard A. McCormick, *Notes on Moral Theology 1965 Through 1980* (Washington, 1981), p. 322. Charles Birch says: Ecological problem is made up of the total impact of "Population x consumption of resources per person x environment deterioration per person. Every item on the impact equation is important." Cf. Charless Birch, "Creation, Technology and Human Survival," *Ecumenical Review* 28 (1976): 70.
8. Bernard Häring, *Free and Faithful in Christ,* Vol. III (New York, 1981), p. 174.
9. The Biblical view of God-man-nature relationship was

vitiated by the Graeco dualism and was thus not faithful to the authentic message of the Bible. Also in view of the ecological crisis, there is need to reconstruct a theology of ecology integrating insights from Eastern theology, process theology, the Teilhardian thinking and complemented by the view God-man-nature in Asian Faiths, Indian cultural traditions.

10. S. Arokiasamy, "Liberation Ethics of Ecology," *Jeevadhara,* 18 (1988): 32-39.

11. Felix Wilfred, *art, cit.,* p. 59.

12. John Carmody, *op. cit.,* p. 31.

13. More than 50% of world's air and water pollution is caused by U.S.A. alone. An average American puts 1500 of pollutants into the atmosphere per year. Cf. R.A. McCormick, *op. cit.,* p. 322.

14. S. Arokiasamy, *art. cit.,* p. 35.

15. *Ibid.,* pp. 32-35.

16. Felix Wilfred, *art. cit.,* p. 67.

17. Elfan Rees, Presentation by the WCC to UN Conference Human Environment, *Anticipation,* No. 11 (July, 1972): 5.

18. Thomas S. Derr, *Ecology and Human Liberation* (Geneva: WCC, 1973), p. 69.

19. George Therukattil, "Doing Moral Theology with the Downtrodden and Integral Human Freedom,"*Louvain Studies* (Fall 1988): 222.

20. Quoted by John Carmody, *op. cit.,* p. 54.

21. John Carmody, *ibid.*

22. F.X. Meehan, *Contemporary Social Spirituality* (New York, 1982), p. 104.

23. John Carmody, *op.cit.,* p. 40.

24. Thomas S. Derr, *op. cit.,* p. 73.

25. Samuel Rayan, "Faith Commitment and Commitment to Justice," *IGNIS* I/1 (1983): 9.

26. Thomas S. Derr, *op. cit.*, p. 68.
27. *Gaudium et Spes,* no. 69.
28. Charles Birch, "Nature, God and Humanity in Ecological Perspective", *Christianity and Criss* 39 (1979): 266.
29. Louis Janssens, "Norms and Priorities in a Love Ethics," *Louvain Studies* 6 (1977): 207-238.
30. G. Maertens, "Ethical Aspects of Modern Technology," in J.A. Selling (ed.), *Personalist Morals* (Leuven, 1988), p. 233.
31. *Ibid.*
32. R.A. McCormick, "Searching for the Consistent Ethic of Lie," in J.A. Selling (ed.), *Personalist Morals* (Leuven, 1988), p. 139.
33. G. Maertens, *art, cit.*, 233.
34. Barry Commoner, "To Survive on the Earth," in Robert. Dish (ed.), *Ecological Conscience* (New Jersey, 1970), p. 123.
35. A.D. Corsen-Finnerty, *World Citizen: Action for Global Justice* (New York, 1982), p. 46.
36. Louis Janssens, *art, cit.*
37. Bernard Haring, *op.cit.*, p. 181.
38. G. Maertens, *art. cit.*, p. 241.
39. Jürgen Moltmann, *op. cit.*, p. 25.
40. E.F. Schumacher, *Good Work* (New York, 1979), pp. 55-61.
41. A.D. Corsen-Finnerty, *op. cit.*, p. 42.
42. Bernard Haring, *op. cit.*, p. 198.
43. *Ibid*, p. 173.
44. Barry Weisbeg, "The Politics of Ecology," in Robert Dish (ed.), *Ecological Conscience* (New Jersey, 1970), p. 151.
45. Bernard Häring, *op. cit.*, p. 184.
46. *Ibid.*, 189.

47. *Ibid.*

48. *Ibid.*, p. 195.

49. Paul Folsom, *And Thou Shalt Die in a Polluted Land* (Missouri, 1971), p. 46.

50. Felix Wilfred, *art, cit.*, p. 68.

51. Charless Birch, *art, cit.*, p. 72.

52. Bernard Häring, *op. cit.*, p. 182.

53. John Carmody, *op. cit.*, 1. 150.

54. George Therukattil, "Social Justice, Primary Requisite for Conversion from Social Sin," *Jeevadhara 17* (1987), 471.

# 4

# ECOLOGICAL CONSCIOUSNESS FROM RELIGIOUS TO TECHNOLOGICAL THOUGHT

In this chapter, we will discuss the emergence of a new form of consciousness, namely ecological consciousness. Though new in our times, this emerging consciousness is not an absolute *novum,* or something that has emerged on our horizons out of the blue. This new consciousness is steeped in the magma of old consciousnesses. Upon reflection one realises that it is distinctive in its form and quite in contrast to the mode of consciousness out of which it has grown, and which it seeks to replace—if only by historical necessity. We will argue that consciousness is not a shapeless could that encompasses humankind from time immemorial but rather that in various epochs of human history it takes distinctive forms, modes or structures. This is also the conclusion of other writers on the subject, such as Carl Custav Jung, Mircea Eliade, Jean Gebser, Ken Wilber, etc. who, in various books and publications, have attempted to reconstruct human consciousness from prehistoric to present times.

Let us underscore the point, namely, that human consciousness, as it functions within cultures, is *highly structured* or finely tuned. We do not have a consciousness as such, but rather we live and function within *specific* modes of consciousness. These modes are culture-bound on one level, and species-bound on another level.

We will not attempt to review the history of the forms of human consciousness in this brief essay. I shall concentrate instead on the dialectics of consciousness which has occurred during the last six centuries, and especially on the dialectics which has occurred during the last few decades.

Some six hundred years ago, religious consciousness was dominant in Western world. This consciousness was wedded to Christian cosmology; was in the jurisdiction of this cosmology. As we shall see later, every form of consciousness is connected with, and a mirror image of a cosmology—by which it is engendered and which it articulates. Religious consciousness and Christian cosmology went hand in hand with each other, co-defined and supported each other.

Guided by religious consciousness, people were adjusted to the universe which was overlooked and steered by God. Their daily affairs were regulated by the awareness of the omnipresence of God, and by the presence of his representatives—the clergy. The point to be emphasised is that the entire field of consciousness of the medieval person was shaped and pervaded by the images of God, by the idea of responsibility of God, and by the desire to be saved and redeemed in God's heaven. Then things began to change. After the messy, turbulent and effervescent Renaissance, a new epoch gathers a momentum in the 17th century. The medieval Christian cosmology was questioned. In the process, religious consciousness was undermined and punctured in many places.

Secularism emerges as a new umbrella under which a new consciousness is being crystallised. We are witnessing the slow emergence of secular consciousness—with its marked opposition to the earlier religious consciousness. This secular consciousness will be articulated in time in the form of technological consciousness. For the time being, (throughout the 17th century) this consciousness is gropping for articulation, for a new distinctive shape.

Humanism of the Renaissance was the first step towards the new non-religious consciousness. Man began to be deemed as the measure of all things. Arrogance of man was in great abundance during the renaissance period. The geniuses of the Renaissance often possessed the ego as big as a mountain. Humanism meant the reassertion of the individual, thus of the ego. When ego is overgrown human arrogance emerges. The arrogance of the Renaissance was actually tempered by the wisdom of the ancients and by the religious constraints which were still very powerful in the 15th and 16th centuries.

The next crucial step in the rise of technological consciousness was the quantification of the cosmos. Only with the advent of the mechanistic world view does the situation change dramatically. The combined efforts of Bacon, Galileo and Descartes (and scores of others, of course) resulted in a new cosmological matrix, a new world view which conceives of the universe as a clock-like mechanism. From this point on, the developments are rapid, far reaching and with staggering consequences.

The new cosmological matrix requires that all phenomena—to be recognised as valid—must be physical in nature or reducible to physical ones. The important relationships about these physical phenomena should be expressed in quantitative laws. Thus, the physical and the quantitative are enshrined. Over time this led to:

— the worship of objective knowledge

— the increasing quantification of all phenomena

— the narrowing of the focus of our vision and our inquiry

— the elimination of the sacred.

A further consequence is the growing process of alienation—a direct result of the process of atomisation and quantification as we split everything into separate atoms, larger wholes are disintegrated. We no longer have a sense of wholeness. But rather a sense of isolation, separation, detachment, in brief—alienation. The psychological alienation is the result of the conceptual alienation; this, we have already established in earlier chapters.

Another consequence of the mechanistic approach to the cosmos is the growing worship of physical power, indeed a sense of intoxication with power, and an obsession with it. A corruption of power, to signify the coercive physical power, is a consequence of mechanistic cosmology—which, with an extraordinary consistency euloizes the physical, the quantitative, the manipulative, the controlling.

It is at this stage that mechanistic cosmology becomes absorbed by the entire culture, that is, Western culture. It then comes to dominate the minds; leads to the emergence of a distinctive form of consciousness, which we call technological consciousness.

We must stress with a due gravity that technology is not to be thought of as a chest of tools—indifferent in themselves; bad or good only according to our use. This is a rather atomistic and naive view of technology, which technology itself attempts to promulgate. At this stage of history, technology is so all pervading that it is a form of consciousness. When we think technology we think "control and manipulation." Technology is a vision of reality—not the use of tools. When we interact with the world via technology, we never think how to be benign and compassionate and loving, but always how to be efficient, controlling, assertive. This attitude of controlling and manipulating is now a part of the mental make-up of Western people.

Another intriguing or shall we say fascinating aspect of technology is that although technological consciousness is supposed to be secular through and through, it contains its own transcendental programme, its own form of divinity. It seeks the divinity of man here on earth—by liberating him from old yokes, giving him dignity and freedom; and by creating him/her in the image of the industrious god who can do everything for himself. Now to seek fulfilment and realisation on Earth, through our own effort, is an admirable project. The trouble begins when we amass too much power with which we destroy natural habitats and by which we become so intoxicated that we forget our place on this planet. In the absence of higher values and some form of wisdom concerning human destiny, the amassing of power is a very dangerous things as it leads to unbridled arrogance and, inevitably, hubris.

Let us now succinctly summarise the characteristics of technological consciousness. When we view its overall structure, its overall mode of operation, technological consciousness reveals itself to be:

1. Objectivising
2. Atomising
3. Alienating
4. Power dominating
5. De-sacralising
6. Geared to the eschatology of consumption.

This last point needs to be elaborated upon. What is

the eschatology of consumption? Eschatology is the subject which concerns itself with the ultimate ends and goals of human life. In the absence of far reaching transcendental goals, while religious and spiritual values have collapsed, consumption has become an imperative of our life, an overall goal, a form of fulfilment, the focus of aspirations.

In its clumsy and indirect way, consumption has become a form of salvation, an eschatology. Consuming new toys, new cornflakes, new cars, new televisions, new computers is not dangerous in itself. What is dangerous, and unhealthy to our psychic lives, is when this process of consumption becomes a kind of religious urge promising happiness, fulfilment, salvation. At this point, technology has become a form of eschatology. In brief, after we have emptied the universe of the sacred, of the spiritual, of intrinsic values; after we declared the physical, the objective, the coldly rational as our new deities; after the traditional cultural patterns have been disintegrated in the miasma of atomistic thinking what has emerged as our new eschatology is the drunken pursuit of power and of stupefying consumption.

What has also emerged is a new image of man—the Faustian man who celebrates his day by seeking gratification here and now. Faustian man maintains that one only lives once, therefore one lives dangerously, at whatever and whosoever's expense, even if it means the ruin of future generations and the destruction of ecological habitats. Faustian man in the human manifestation of the rapacious technology. Faustian man is a symbolic acknowledgement of the ascent of naked power and the simultaneous waning of human spirituality.

While analysing the fallout of technological consciousness, we are not forgetting the other side of the coin, namely, technology has been a noble dream that has turned out sour, that the mechanistic world view was once a ladder to freedom—from the constraints and oppressions of religious consciousness. Nor do we wish to question the obviously beneficial aspects of technology—comfort, the rise of the material standard of living, the sense of freedom of movement (if only illusory), the elimination of contagious diseases, an easy way of plumbing man of our problems.

Yet as we approach the 21st century, technological consciousness appears increasingly as a menace with its ruthless efficiency, its uncontrolled increasing of power, and its lack of any compassionate accounting. Technological consciousness simply does not add up. It produces too many diseases of its own. And the price for comfort and other amenities is simply not worth paying. This is the realisation to which we, as a society, have come to only gradually.

Technology is a story of success which has been so stupendous that it has become our nightmare. The pendulum has swung far too much in the direction opposite to religious consciousness. We thus seek to re-establish a new balance, and in the process, we attempt to balance our own lives.

Ecological consciousness which is outlined in this essay is a synthesis. Religious consciousness was the thesis. Technological consciousness was the anti-thesis. Ecological consciousness is the synthesis as it marks a return to the spiritual without submitting to religious orthodoxies and the religious dogma; and as it seeks social amelioration and justice for all without worshipping physical power and without celebrating the aggressive nature of the human person.

We are not claiming that economical consciousness has arrived, is fully articulated, and comfortably dwells in us. Rather we are suggesting that in reaching out, we bring to fruition that which is dormant and which wants to be awakened and articulated. Our projection is part of the articulation, is an essential part of the creation of ecological consciousness.

## THE RISE OF ECOLOGICAL CONSCIOUSNESS

All modes of consciousness are rooted in history and have a historical character. They come at a certain point of history and disappear or are profoundly modified, at another point of history. Thus consciousness is determined by history. But, in its turn, it determines history. In our times, we have been witnessing *the greening of consciousness.* Simultaneous to this process of greening, another process is going on: the process of the greening of world religions, although the latter is less easily perceptible. Let us be aware that from the time of the Congress of Assisi (1986) where five major religions were represented, the ecological interpretation of world religions has acquired an important momentum.

The forerunner of ecological consciousness was the Ecology Movement, on the one hand, and various schools of Humanistic psychology, on the other hand. In their respective ways, they were against the temper of the mechanistic age. Both have emphasised wholism and the irreducibility of large complex wholes to their underpinning components: ecological habits and human persons. Both these movements were a challenge thrown to the rationality of the mechanistic system. Both movements professed a new type of wholistic rationality.

Moreover, in a certain sense both these movements possessed a religious flavour. They offered not only new intellectual vistas but a form of liberation. This liberation, although not always explicit, was meant to give us freedom from the deterministic and mechanistic shackles. Wholism, which both movements emphasised, was the first step to liberation.

What are the chief characteristics of ecological consciousness? We shall enumerate six such characteristics and contrast them with the respective ones of technological consciousness. We shall not claim that these six characteristic completely define the scope and nature of ecological consciousness. We have to simplify. To understand is to simplify.

| *Ecological Consciousness* | | *Technological Consciousness* |
|---|---|---|
| wholistic | vs. | atomistic |
| qualitative | vs. | quantitative |
| spiritual | vs. | secular |
| reverential | vs. | objective |
| evolutionary | vs. | mechanistic |
| participatory | vs. | alienating |

A more appropriate form of expressing the nature of ecological consciousness would be through a mandala as each of its characteristics is feeding into each other and feeding on each other; co-define each other

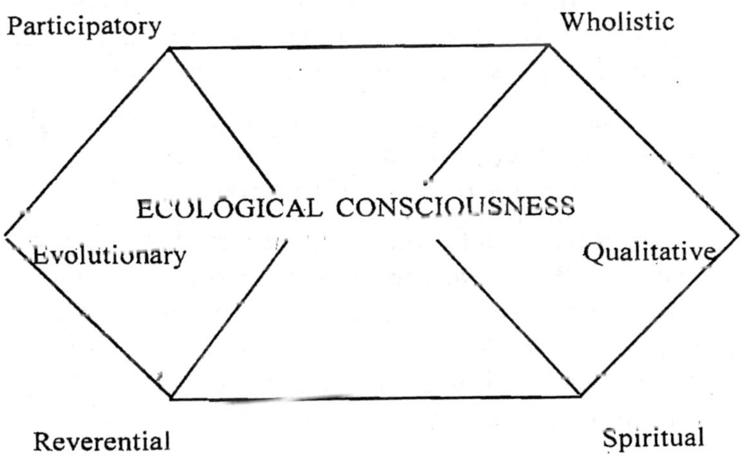

Let us now discuss these characteristics and see what it all means for our own lives and for our perception of the universe.

That a healthy and complete human person is a micro-universe which is wholistic and qualitative—there no one can doubt. That a human being who seeks meaning beyond the triviality of consumption is on some kind of spiritual path—there can be no doubt as well. The quest of meaning is a spiritual quest.

As for the attitude of reverence, any person who truly respects others, and truly appreciates the amazing alchemy of the universe cannot be but reverential vis a vis the awesome spectacle of creation. Thus, reverence is an aspect of seeing and understanding the universe—in depth and with a true appreciation.

To live in grace is to be in a continuous mode of reverential understanding. To live in grace is to think reverentially.[1] To live in grace is to walk in beauty—as a song of American Indians proclaims.

Reverential understanding as well as the reverential attitude are not new creations of ecological consciousness. They have existed in traditional cultures and religions for a long time. We are only articulating them de novo.

The natural condition of the human person who is alive is to be enchanted by the world. Reverence is acknowledgement of this enchantment.

A rational skeptic may respond at this point: "It is all very well to treat the human person in a wholistic, reverential and spiritual manner—particularly if you are a humanistic softie, but how can we demonstrate, how can we *prove* that the physical universe is wholistic, reverential, spiritual?"

We are rather fortunate that at this juncture of history,

science at the cutting edge provides us with an unambiguous evidence of the wholistic nature of the universe. David Bohm's book *Wholeness and the Implicate Order* is a case in point. Bohm's views are rather well known so we shall not belabour them here. Nowadays, it can be consistently and legitimately maintained that the physical cosmos, in its stupendous evolution is one interconnected whole of which each part partakes in the glory of the whole, and somehow knows of the existence of all others. Bohm is not the only physicist who assumes that by its very nature the universe is wholistic since we all come from the same fireball.

Equally significant, if not more striking is the evidence emerging from astrophysicists who have been wondering why the structure of the universe is as it is. The balances in the overall structure of the universe are exquisitely tuned. It is now clear that if the universe had different density, dimensions and time-span, life could not have emerged. It has been calculated recently that if the composition of some of the basic elements in the universe would have been altered by as little as 10%, the whole fabric of the cosmos would not be able to generate life. Thus, this laboratory, called the cosmos, seems to have been uniquely set to generate life. Freeman Dyson writes: "As we look out into the universe and identify the many accidents of physics and astronomy that have worked together to our benefit it almost seems as if the universe in some sense have known that we were coming."

The conclusion has emerged slowly—Why is the universe as it is? Because we are here. The universe is as it is because it meant to generate intelligent life. This insight led to the formulation of the Anthropic Principle (in 1981) which proclaims that the universe is home for man. Its composition, structure and dynamics were exquisitely balanced to enable life to emerge. As we search deeper and deeper into the underlying structure of the cosmic evolution, we are more and more convinced that "The coincidences" may not have been so coincidental.

The universe has brought life to celebrate itself. We are part of its glory. We neither deny our special place nor are unduly arrogant about it. It has just happend that we are part of the flowering of the universe. To deny this special pace to homo sapiens in the name of ideology of anti-anthropocentrism is a folly based on a new form of misanthropy; is, in fact, a form of inverted arrrogance.

From the standpoint of ecological consciousness, the Anthropic principle reaffirms the wholeness and unity of it all, and reassures us that we are legitimate dwellers of the cosmos, not some kind of cosmic freaks. In one sense, we are the justification of the nature of the universe: the tremendous cosmic changes have worked out to create life endowed with intelligence. On another level, the conception of the universe as home for man implies that we are its custodians, responsible for all there is, including our own fate.

## FROM THE ANTHROPIC PRINCIPLE TO REVERENCE FOR LIFE

Our rational skeptic is now nodding his head and says: "Although we can somehow agree that the universe is wholistic in nature, how can we *prove* that it is reverential and spiritual in nature? Where is the evidence for that?" We shall respond. The evidence is in our mind. To think beautifully about the universe is to think reverentially of it, is to behold it with reverence. The skeptic may object that this is not the kind of evidence he has expected. At this point, we shall say that the skeptic has not grasped the subtlety or our argument. We are not saying that reverence is one of the *physical* aspects of the universe, to be found among other physical attributes. We have never committed ourselves to such a proposition. We are saying rather that reference is an attribute of the mind. If we behold the universe reverentially, we dwell in a reverential universe. If we behold the universe mechanistically, we dwell in a mechanistic universe.

The mechanistic nature of the universe can be attested to by the existence of present mechanistic science and technology. How can the reverential nature of the universe be attested to—the skeptic may further press. By the existence of reverential science and reverential technology—although these two have not been much developed. It would not be true however to say that they do not exist.

Various systems of *yoga* represent reverential technology. And, we should not raise an eyebrow or dismiss the idea because we are used to thinking about technology as physical tools engaging us with the physical world. Our interactions with the universe are numerous and subtle. Any tool or technique which engages us with the universe is a form of technology. Yoga systems are the techniques of the soul. Prayer is a form of technology—if it does something to you and facilitites your interactions with the universe—divine or otherwise.

Let us reflect on the nature of cosmology. Cosmology makes assumptions about the universe *in toto*. And then, acting on these assumptions, it finds in the universe what it has assumed to be the case. Such has been the story of most historically known cosmologies. Cosmologies *do not prove* the existence of this or that attribute which they assume about the cosmos. They proceed as if this attribute was inherent in the structure of the universe. And, then they build large patterns of perception and knowledge which vindicate the assumed existence of a given attribute.

*Nothing reveals itself in the cosmos unless we assume*

of divinity as the mind (*nous*) is the creator of all orders including the divine or spiritual order. Sacredness is an attribute of the mind, not an attribute of the cosmos. Only when we approach the universe with the reverential attitude the behold it by the mind that is sacred, we find the universe sacred.

Now whatever we have said about reverence, applies to spirituality (the fourth element in our mandala). If we assume that the universe contains nothing but dumb matter[2] and the only way to know this universe is through physical knowledge, we shall obviously not find any traces of spirituality in this universe because our assumptions and our language have ruled it out.

If we assume, on the other hand, that the universe is spiritually alive and that we are spiritual agents, and if we act on these assumptions, we shall find ample evidence that the universe is transphysical and transbiological or simply spiritual, as so many people of spiritual inclinations have found in history.

## OF EVOLUTION AND HOPE

Let us briefly discuss the last two characteristics, evolutionary and participatory aspects of ecological consciousness. The beauty of our universe is inherently connected with the beauty of evolution—this awesome process within which amazingly creative forces have co-operated to bring about one miracle of life after another. The French philosopher Henri Bergson has captured the essence of this process by coining the phrase "creative evolution." His compatriot, Pierre Teilhard de Chardin was a rhapsodic bard singing countless hymns to creative evolution. Teilhard did much to convince us that evolution is not a dreary Darwinian process of slugging from one chance to another chance, from one incomprehensible necessity to another.[3] No, evolution is not a stupid and chancy process of stumbling upon one beneficial variation after another. Evolution is so exquisite in its mode of operation that it could be called divine. I, myself, have no difficulty in accepting the idea that God is evolution; and evolution is God.

In brief, ecological consciousness is evolutionary par excellence. Why is the acceptance of creative evolution important to the overall structure of ecological consciousness? Why is the right comprehension of evolution important to our sense of the future and to our destiny?

For three reasons, a sensitive reading of evolution informs us that the universe is quintessentially unfinished, that we are essentially unfinished. From, this is clearly follows that we shall further unfold as the cosmos will unfold. Another important conclusion that follows is that we have somewhere to go, that we have a stupendous future in front of us. We are still toddlers in the cosmic playpen. we shall mature, we shall take our destiny more resolutely into our hands; we shall become less stupid, less vulgar,less consumptive; more frugal and more wise. This is what an intelligent reading of evolution informs us about.

Secondly, an intelligent reading of evolution informs us that evolution is a *divinising* agent, that it transforms matter into spirit. Although, some people claim that consciousness must have been present in matter and the entire universe from the beginning (otherwise how could it come to exist)—I do not accept this view.

Consciousness is *emergent.* It came to exist at a certain point of the development of matter. Even if we could agree that consciousness could have been there — deeply hidden in the inner layers of matter...waiting to be released—I would still argue that this process of releasing it from the bondage of matter was so extraordinary and would possess an altogether different structure of mind, our pictures of the world, and all our understanding of it, would be different. It is in this sense that our consciousness is species-bound.

We never describe the cosmos as it is. We always partake in what we describe. Our mind invariably and tirelessly elicits (through the various faculties and sensitivities we possess) from the amorphous primordial givens of the universe. Given is never given as such; it is always mediated, moulded, shaped and determined by the mind.

We have already discussed the importance of the participatory mind while we introduced the noetic conception of divinity, and while we have pointed out that the mind (or human consciousness) is the creator of all orders, including the spiritual order. We also have mentioned that if we assume the mechanistic nature of the universe, we shall read the universe in a mechanistic way. On the other hand, if we assume the reverential nature of the universe, we shall read it reverentially and interact with it reverentially. These notions clearly imply the participatory, co-creative nature of the mind which co-creates with the universe.

Our mandala is now completed. We have articulated six essential characteristics of ecological consciousness. This consciousness is wholistic, qualitative, spiritual, reverential, evolutionary and participatory. These characteristics from one coherent whole.They partake in the making of Eco-cosmology, on the one hand; they are consequences of this cosmology—on the other hand.

The process of the Greening of Consciousness is going on. But a word of caution would be advisable. This process will not happen automatically. It will require great energy and will power on our part. Nor will evolution proceed automatically and effortlessly to its happy end—the Omega Point—without our help. We are evolution conscious of itself. That is true. this realisation imposes on us an enormous and a beautiful burden—taking the responsibility for all there is, for our own future and for their future of evolution.

It is sometimes assumed that evolution is working with swiftness and necessity and will deliver us to some promised land no matter what. It is also assumed that evolution is making such stupendous breakthroughs that it is only a matter of years when the bulk of humanity becomes elightened, and we shall be all on the right path. These views we consider altogether much too optimistic.

To change, the human consciousness presently stuck in the mechanistic mores will require a painstaking effort. I will require some kind of revolution. Indeed this is how Ionesco envisages revoluation: "Revolution is a change in the state of consciousness." Incidentally, revolutions which did not work, including the Soviet Revolution, are the ones which failed to create a new consciousness.

Now let us consider another subtle but important point. Why is a change of consciousness so difficult on the individual level? Because the organism sees it as a challenge to its identity. Consciousness is identity. When our consciousness is in the process of change, we are in the process of change. We are comfortable in our old niches—whatever they are. Old niches are tantamount to stability. New consciousness implies instability to begin with, even if it leads to liberation and new freedom in the long run.

How do we change our consciousness or at least a part of it? Compassion, reverence and wholistic thinking are not given to us *Deus ex Machina* but require diligent systematic work. I call these special exercises, which lead to the acquisition of ecological consciousness, Eco-yoga. This is nct the place to explain the principles of Eco-yoga, particularly as any yoga is to be practices not talked about. Spiritual practices of old venerable traditions may be of considerable help. Many of the logic systems presently available may be of considerable help too—if discriminately practised.

Each of us will need to find his/her own special exercises which will lead to a change of consciousness—if such a change is deemed desirable. Nobody can force anybody to acquire a form of consciousness which is not deemed desirable. Ecological consciousness offers itself as a desirable candidate because it promises to solve fundamental problems of our times at their root causes. Besides, it gives us the luxury of knowing that the universe is home for humans and we are its legitimate dwellers, perhaps even exalted ones.

## SUMMARY

A change in the nature of consciousness signifies a change in the description of the world. At such a point of change we are here at present. Through, the acquisition of ecological consciousness we are changing the nature of our perception, we are changing the nature of our knowledge, we are changing the nature of external reality. As we filter and sculpt reality differently, so we receive it differently.

The relationship between ecological consciousness and Eco-cosmology is a very close one; one co-defines the other. One needs the other for its existence. It may be said that ecological consciousness represents the interiorisation of the principles of Eco-cosmology. Eco-cosmology, on the other hand, needs ecological consciousness as its articulator and as its representative in the world of consciousness. Cosmology does not exist by itself. It comes to existence when it is articulated by the appropriate modes of consciousness. Those appropriate modes of consciousness we have to acquire individually—by working on overselves. Only then will ecological consciousness become a social reality.

Ecological consciousness is finding as structure of sustainable and compassionate beauty that encompasses it all.

## REFERENCES

1. For further discussion of reverential thinking see the chapter on the subject in my book: *The Theatre of the Mind.* 1985.
2. The more sensitive the mind, the more sensitive matter becomes. Some scholars talk about the sensitivity of matter. But this sensitivity reveals itself only to sensitive minds—sensitive enough to assume that matter is not this "dumb stuff out there."
3. I am referring here to Jacques Monod's book:*Chance and Necessity,* which attempted to reinforce the materialist world view and the Darwinian concept of evolution.
4. I have been giving workshops on Eco-yoga, as a part of acquiring ecological consciousness, during summer time on the Island of Tharsos in Northern Greece.

# 5

# TRANSCENDING ECOLOGY: CHRISTIAN AND MARXIST APPROACH

Christianity emerged in order to solve some important problems which the crumbling Roman Empire could not solve. The disintegration of social and individual life in the fourth and fifth century Rome was at least partly caused by the fact that the centre did not hold. There was no centre any more; no matrix to support man's spiritual life and sustain man's deeper aspirations and visions. Christianity created a new centre, and with it a new spiritual matrix. Out of this centre and this matrix grew a new civilisation, a new social order, and a new conception of man. For a millennium, the Christian paradigm held well. From the 16th century on, it has been holding less well—new problems emerged to which Christianity could not provide an adequate response. As, the result new philosophies and ideologies were created to face and solve specific problems of their epochs.

Marxism emerged in the 19th century to solve some important problems of this century. In the epoch of growing prosperity, of enlightened reason and progress triumphant, it was an insult to human dignity and human reason itself that so many people would live in the squalor produced by the Satanic Mills of the 19th century industrial revolution—exploited ruthlessly and shamelessly. Social justice was an imperative following from the premises of Enlightened Reason. At least Marx thought so. And, he set himself to rectify the undesirable social conditions which, thinking: and conversely, on our capacity to integrate analytical thinking into the framework of sustainable values.

In so many traditional societies, the Buddhist especially, there is a clear awareness that right thinking, and right attitudes lead to right action. I will let the poet T.S. Eliot make a main point for me:

> *A wrong attitude towards nature implies somewhere, a wrong attitudes toward God, and the consequence is an inevitable doom.*

## CONSERVATION IS AN ETHICAL ACT

When excessive specialisation claims larger and larger parts of our minds, conservationists may be the only true guardians of the whole earth. Insofar as they are part of the scientific community, they are under the strictures of the accepted methodological principles. However, insofar as they are the true guardians of the whole earth, they may have to overcome and transcent some of these strictures.

What is conservation? Conservation is an act of caring to the point of fighting for what you consider important to preserve—even against considerable odds. When the world is shrinking and the earth is devoured by expanding populations and expanding concrete surfaces, how can we establish and maintain a right and rational strategy for preservation? The answer is not by treating habitats we wish to preserve, as choice pieces of property, but by regarding them as something of *intrinsic value.*

We must justify eco-habitats to be of intrinsic value, and not as pieces of commodity to be evaluated by market strategies, by realising that conservation is an ethical act, which is based on the premise that we are preserving the richness and uniqueness of life itself.

This of course assumes the recognition of the unity of life, of which we are a part. This unity of life must be viewed not in the trivial sense, as paying lip service to the idea that we are all interconnected, but in a deeper sense, by acknowledging reverence for all life, and by making out stewardship of life a principle of paramount important.

The reconciliation of these two seemingly incompatible ethics: of objectivity and of caring, though difficult, is both necessary and possible. We must simply look at the criteria of objectivity for deferent kinds of systems. The objectivity of the very small (sub-atomic particles) and the very large (galaxies and black holes) are not the same thing. The objectivity of the very simple and the objectivity of the very complex are again governed by different criteria. The objectivity of systems without a time dimension and of systems evolving in time are different again. We must therefore not apply one kind of objectivity (characteristic of simple timeless systems, which we analyse in chemistry for instance) to all other systems: of different complexity, different dimensions, of enormous number of variables, and in which time is a most important factor.

When properly analysed, we shall find as many different types of objectivity as there are disciplines examining different aspects of life—in or of the universe. An objective way of looking at the life process, when we attempt to understand life as life, is one which must enable us to understand the diversity of the forms of life within the unity of all life. This act of deeper understanding is based on the faculty of empathy, or identification with other forms of life, which is, at the same time, an act of caring. This is a moral act, in a sense an act of participation in life. Therefore, forget about Jacques Monod (1971) and his magnum opus *Chance and Necessity,* for what he promotes is not understanding life but reducing life to its chemical components. Yes, we shall not deny that there is some light that comes from the study of molecular structures of living organisms. But there is a different principle of objectivity in action when we examine molecular structures in the laboratory, on the one hand, and when we examine life in evolution, on the other hand.

We have deliberately chosen the case of the objectivity of chemistry, as perhaps the most confining one, for physics nowadays is quite a different matter. Already with Einstein (relativity), Heisenberg (uncertainty) and Bohr (quantum theory) enormous difficulties has arisen in applying the criteria of objectivity characteristic for the Newtonian system. In recent years, the New Physics has been bursting in new and unexpected directions, to the point that the principle of objectivity (in the old-fashioned Newtonian sense) is considered an antiquated thing of the past.

The proponents of the New Physics are telling us that we live and think in the *participatory universe.* We cannot be objective, and never are objective, in the strict old-fashioned sense (even in molecular biology). We are always co-creators participators.

The new concept of objectivity, if we are still to preserve the term, is one of the participatory process, and this concept is excellently suited to our new ethics of conservation. Conservation as an ethical act is now well justified by the strategies and methodologies of the New Physics and, especially, its concept of participation.

Before we attempt to outline some principles specific to eco-ethics, let me share with you a couple of fragments of the great speech of the Red Indian Chief Seattle, delivered in 1854. In this speech, in a poetic and metaphorical way, he captures the spirit of ecoethics. It is this spirit that we must bear in mind when we contemplate conservation as an ethical act, and other principles of eco-ethics. Chief Seattle said: "The Great Chief in Washington sends word that he wishes to buy our land... How can you buy or sell the sky, the warmth of the land? The idea is strange to us. If we do not own the freshness of the air and the sparkle of the water, how can you buy them? Every part of this earth is sacred to my people. Every shining pine needle, every sandy shore, every mist in the dark woods, ever clearing and humming insect is holy in memory and experience of my people. The sap which courses through the trees carries the memories of the red man... Teach your children what we have taught our children: that the earth his our mother. Whatever befalls the earth befalls the sons of the earth. If men spit upon the ground they spit upon themselves."

This we know. The earth does not belong to men; man belongs to earth. This we know. All things are connected like the blood which unites one family. All things are connected. Whatever befalls the earth befalls the sons of the earth. Man did not weave the web of life, he is merely a strand in it. Whatever he does to the web, he does to himself.

We would not have designed all the various conservation strategies, if we did not feel responsible for the survival and well-being of eco-habitats and whole environments. Responsibility is a moral category. No scientific description of any habitat can make us responsible for it sunless and until we feel responsible for it. Feeling and being responsible for environments and natural resources is being responsible for the world larger than one self and also being responsible to future generations; of humans and non-humans as well. Our responsibility is effective if we inspire others to be responsible. Responsibility is contagious, for it has a moral force behind it.

However, we have a problem here. Our culture, and by this we mean Western culture, seems to discourage us from acknowledging values as a legitimate part of our scientific and cognitive endeavours. Thus in spite of our better knowledge, we often deny or suppress value considerations though we deep down know of their importance. This is the result of the positivist philosophy that has crept into our academia and our thinking and which tells us that values do not belong to the realm of rational cognition and must be therefore eliminated from man's rational discourse, which is of course nonsense when you come to reflect upon it. Yet, the atrophy of values is a fact. We are all-constrained by the positivist strait jacket.

What we are at present pursuing is not realism but the worst kind of fiction which is leading us to our own extinction . If so, then we shall have to break the conspiracy of silence, for this conspiracy does not serve any form of realism. Instead it is aiding and abetting a destructive fiction. Let us also reflect that if to know the truth, and stand by it, is to appear romantic, then there is something wrong with the whole structure of society.

The responsibility for environments and for their right conservation is our moral obligation if we are to leave behind any meaningful heritage for future generations; and it is our responsibility also to instill the sense of responsibility in each other. The World Charter for Nature (Point 8), proposed at the UN in June 1980 argues in a similar vein: Man must acquire the knowledge to maintain and enhance his ability to use resources in a manner which benefits present and future generations without lasting injury to nature. Man can be in harmony with nature if the human community acts as a steward for nature in the interests of future generations.

The most effective strategies for conservation and maintenance of natural resources are not through mere economic calculations, but through the promotion of right eco-values. This point may appear debatable, at least in some quarters in which cost-benefit analysis is elevated to a supreme criterion. Let us therefore examine the situation more closely. The cost benefit analysis itself is an expression of a value judgement for it tells us that we should value economic renumeration over everything else. This is an important point of remember: we are encouraged to abandon traditional values in favour of the economic value, in favour of reaping profit in the short run. There is clearly a value judgement behind seemingly objective cost-benefit analysis. As conservationists and comprehensive thinkers dealing with global strategies, we must resist this kind of subversion of our thinking based on a value which—as we have seen over and again—is so inadequate in the long run. If we are so clever that we can exploit everything to our advantage, then we must not be so stupid as to cut off the branch on which we sit.

Yet, development is important, particularly for the Third World nations. Incidentally, we often have the feeling that the idea of development has been made into a fetish in the developing countries. What 'progress' signified in the West some 20 years ago—the sacred cow beyond criticism and shadow of any doubt—Development came to signify in the Third World nations. Be it as it may, we have to find a way of reconciling conservation with development. Now, let us reflect and realise that the every formulation of our dilemma is slanted in favour of development. For reconciling conservation with development almost invariably means constraining conservation to the demands of development.

We should have the courage and wisdom to look at the total equation, at the whole panorama of man's culture, including his spiritual needs. What is specifically mean is beautifully rendered in the book by Colin Turnbull (1962) in which the lonely African thus speaks: we have tried hard to understand the white man and his ways, but we can see only harm. What happiness have they brought us? They have given us a road we did not need, a road that brings more and more foreigners and enemies into our midst, causing trouble, mankind our women unclean, forcing us to a way of life that is not ours, planting crops we do not want, doing slave's work. At least the BaNgwana left us our beliefs, but the white man even wants to steal these from us. He sends us missions to destroy our belief and to teach our children to recite fine-sounding words: but they are words we believe in any way, most of them. And we live according to our beliefs, which is more than the while man does.

Is the lonely African with his values going to be melted in the melting pot of one homogenised culture?

In this context, let me suggest that all ethics have always been in the service of conservation and preservation. All ethics wish to preserve and conceive the sanctity of the human being, his wholeness, his integrity. The virtues that various systems of ethics advocate and wish us to uphold always serve some programme of conservation. Here, therefore is another argument for our earlier claim that conservation should be seen as an ethical principle. For all ethic is conservation; and all conservation is ethic.

How do we preserve the integrity of the people still

adhering to traditional life-styles in our times of change and development? Perhaps by maintaining that not only every culture is of intrinsic value and should be spared but also that every form of life is unique, warranting respect, regardless of its present worth to man. This principle is so expressed in Point 8 of the UN World Charter for Nature, which we have already mentioned.

Now, in order to treat every form of life as unique, and as a value in itself, we must develop certain attitudes, certain frames of mind, in short, values specific to eco-ethics which would inspire right behaviour and make us take seriously our intrinsic evaluations of eco-habitats. When we look deeper into the attitudes that are required us to treat every form of life as unique: when we inquire what modes of thinking would be required to respond adequately to the demands of the heritage of whole life, then we must come to the conclusion that no form of objective thinking will ever suffice. Instead we must develop and implement *reverence for life.*

Reverence for life is not you and me *talking* about it. If it is to become a reality we must change the modes of our thinking. We must develop what we should like to term as *reverential thinking,* which is quite a different kind of thinking from one that is thrust upon us in present schools and academia. The 'official' thinking is an objective one which often goes against the grain of reverential thinking and the principle of reverence of life itself.

To think reverentially is first of all to recognise human *life* as an intrinsic value; it is to recognise *love* as an essential and indispensable modality of human existence; it is to recognise *creative thinking* as an inalienable part of human nature; is it to recognise joy as an inherent part of daily living; is it to recognise the *brotherhood of all beings* as the basis of our epistemological paradigm. We have once again returned to the idea of Reverence for

Life which is all providing and crucial for all aspects of the Ecological World View.

Aldo Leopold (1949) was clearly anticipating eco-ethic, which he called a land ethic, in his memorable book *A Sand County Almanac,* He wrote: All ethics so far evolved rest upon a single premise: that the individual is a member of a community of interdependent parts. His instincts prompt him to compete for his place in that community, but his ethics prompt him to cooperate. The land ethic simply enlarges the boundaries of the community to include soils, waters, plants, and animals, or collectively, the land.

The enlargement of ethic so that it includes as part of the global eco-community soils and waters, plants and animals, rocks and other beings, brings us to the idea of eco-justice. This idea has been of late pursued by churches which are becoming more and more ecologically aware. There are two simultaneous movements converging; the churches slowly enlarge their conscience to include justice for all beings in this universe; so that we read in one of the pamphlets, on eco-justice (produced by American Baptist National Ministries), that 'Humanity and nature can come together in eco-justice.' The pamphlet goes on to explain that eco-justice means joining together concerns for ecology and justice. It expresses the age-old dream that 'Earth shall be fair, and all her people wise.' The other movement is the ecology movement which is slowly gravitating toward the idea that we shall need a spiritual basis to make an ultimate sense of our ecological concerns. An eco-theology is thus is the making.

What reverence for life means in terms of our thinking, perception and attitudes toward all beings is beautifully expressed in the passages of the Chief Seattle we have already quoted. Traditional life-styles of many cultures provide eloquent evidence that reverential thinking is not only possible but has been practised on a large scale. The strength and sustainability of many traditional cultures and traditional life-styles consisted precisely in enshrining the attitude or reverence for life as a modus of daily life. In brief, reverence of life as an essential value of eco-ethics and eco-culture, must not be seen as a luxury but as an endorsement of a way of looking at life.

Organisms not only live in large habitats but they are also a product of evolution. Evolution itself is a value. It is within the matrix of evolution that everything happens. In both human realities and ecological habitats, there are always conflicts between different demands of life. Reverence for life should be maintained as much as possible. But we cannot extend it to all living forms *all the time.* In times of conflict how should we decide which form of life is more important, if we claim that each form of life is unique? we suggest that we use the criterion of evolution: the more exquisite the form and the more accomplished the product of evolution it represents, the more worthy it is of preserving. In my book *Eco-Philosophy: Designing New Tactics for Living,* we have outlined the evolutionary imperative and its values. If we accept evolution as continually transcending itself, in other words as emergent; and if we accept ourselves as a part of the evolutionary process, the question is, what values can we extract from the evolutionary process and how can these values, if judiciously applied, shape our attitude and behaviour? we suggest that the following precepts can be extricated from intelligent reading of evolution:

* behave in such a way as to preserve and enhance the unfolding of evolution and all its riches;
* behave in such a way as to preserve and enhance

life, which is a necessary condition for carrying on evolution;

* behave in such a way as to preserve and enhance the eco-system, which is a necessary condition for further enhancement of life and consciousness;
* behave in such a way as to preserve and enhance the capacities which are the highest developed form of the evolved universe: consciousness, creativeness, compassion;
* behave in such a way as to preserve and enhance human life which is the vessel in which the most precious achievements of evolution are contained.

These five characteristics of the evolutionary imperative are only variations on the same theme. They all follow from the first formulation. However, the articulation of the first formulation makes us aware and gives us a guidance on how to act in times of conflict. From the standpoint of the evolutionary imperative, if we have to take the life of a human being or a mosquito, we should not hesitate but take the life of a mosquito—for it is a far less accomplished product of evolution. We have always known this by instinct. The evolutionary imperative justifies as a right moral principle that which we have acted upon through our instincts.

My evolutionary imperative has been challenged by Deep Ecology proponents as heedlessly anthropocentric; indeed as spelling out a new version of the old-fashioned concept of the superiority of man over all other beings, and as giving a *Carte blanche* for the exploitation of other beings for the benefit of man. Yet, we wish to submit that this old-fashioned anthropocentrism cannot be found in my imperative. What can be found in my imperative is an awareness that ethical choices are choices of values, and when we are faced with a choice: whether to take the life of a mosquito or the life of a human being. We choose (on the basis of the evolutionary criterion) the life of a mosquito. What other criteria are there? Total egalitarianism, according to which every form of being has an absolutely equal right, is a nonsense from the human point of view, and we must emphasise—that this is the only point of view we have: *human,* even when we wish to argue against this point!

Total egalitarianism is also against the *modus operandi* of nature; and of the whole of evolution—as we understand it. Do the proponents of Deep Ecology wish to say that they know better than nature and evolution? And what is their special mandate that enables them to cast out all other views which disagree with theirs—when the subject is so intricate, complex and difficulty? Moreover, from the standpoint of the Third World nations, the gospel of return to original natural conditions of the hunter gatherer societies, whereby we shall have to eliminate some 80% of the world's population—this concept means a systematic genocide. For where do we start 'eliminating' the overabundant populations (may we ask Deep Ecology people) if not from so-called over populated countries? Thus on one interpretation Deep Ecology is spelling out a version of population fascism; unintended by Deep Ecology people no doubt, but embedded in some premises of Deep Ecology nevertheless.

Self-reliance of individuals, nations and cultures is an ethical imperative so that the idea of responsibility (of individuals and nations) does not become meaningless: Eco-Justice demands that we not only assume our responsibilities, but also that we have the scope and the possibility to do so. Self-reliance, the capacity to be your own master, thus appears to be one of the values indispensable for eco-justice. This refers to the self-reliance of individuals, who must not be dwarfed by the machine, bureaucracy, or some other forces that crush the will of individuals for themselves and their eco-habitats; and this also refers to the self-reliance of nations which must be able to sustain themselves and not be so dependent on and so determined by other nations that the will and interests of these other nations will make mockery of self-management and indeed the sovereignty of dependent nations.

Self-reliance of individuals and nations is one side of the coin; ecological diversity is the other side of this coin. In order to do justice of the variety of lands, climates, circumstances and traditions we have to cultivate diversity in ecological, agricultural (as well as cultural) terms. For this diversity is the basis of self-reliance; and vice versa; self-reliance in so vastly varied circumstances of our globe, and within different traditions simply means encouraging and maintaining diversity. Diversity means heterogeneity; means the opposite of homogeneity. Homogeneity profits central economies, high-tech homogeneity. Heterogeneity profits local people; enables them to be self-reliant; ultimately enables them to be responsible and good stewards.

We, therefore, have a cluster of three concepts: self-reliance, diversity and heterogeneity, which do not seem at the first sight to belong to the realm of ethics, and yet on a closer inspection have clearly ethical implications; are in fact a part of eco-ethic.

To put the matter in a crude and simplified manner: it is immoral to allow homogeneity to prevail.

Already in the early 1970s, the Polish writer Alexandrowicz wrote a penetrating essay entitled *Ecological Conscience* in which he anticipated the coming of Eco-ethics. In the late 1970s and early 1980s, the awareness that our ethic must be ecologically based has been emerging from various quarters , among which the churches seem to be taking more and more decisive initiative—calling not only for eco-justice ('Humanity and nature can come together in eco-justice'), but also calling for enacting the 11th Commandment:

> The earth is the Lord's and the fullness thereof: Thou shall not despoil the earth, nor destroy the life thereon.

It may be said without exaggeration that unless we develop and implement eco-ethics and based on some intrinsive values, our work as conservationists will be superficial and ineffective. It is to be hoped that working conservationists in all lands and cultures will not shy from value considerations, but will consciously and deliberately include them in their work because these values are the hidden springs of our action and thought.

# 6

# ECO-PHILOSOPHY OF BUDDHISM

It is a long long way from academic analytical Western philosophy to Buddhism or Zen Buddhism. Yet, many of us have travelled this road in recent years. Some travelled it by jumping out from the wagon of Western philosophy; often by rejection of Western rationality and Western tradition—straight into the gentle embrace of oriental mysticism, be it Hinduism or Buddhism or Zen Buddhism. Some have travelled this road through an internal reconstruction of Western rationality itself; by attempting to work out a new rational framework—not by rejecting rationality but by creatively transforming it. I myself belong to the latter group.

The Odyssey of western mind is endlessly fascinating. I will try to reconstruct part of this Odyssey as I have experienced it. In the process, I will show how an attempt to reconstruct Western philosophical tradition inadvertently (or may be inevitably) leads back to venerable ancient spiritual doctrines of the past.

I was raised in a typically rational tradition of Western philosophy. M.A in logic from Warsaw University which was at the time one of the strongholds of analytical phisolophy. Such was the reputation of the Polish School that when I found myself at Oxford, I did not have much difficulty in obtaining a scholarship from one of the prestigious Oxford colleges.

In due time, I received a D.Phil. from New College for a dissertation in analytical philosophy. All was well. I have a coveted degree from the most prestitious university in the world. Philosophy was at its acme at Oxford at the time. Except that not all was well. Already at Oxford I became increasingly disenchanted with the aridity of Oxford linguistic philosophy. It was so brilliant, clever—yet empty inside. A very fascinating game to play—if you wanted to devote your life to that kind of game. Many have done so. Not because they are obtuse. But rather because they are very cleaver. They go on through life pretending that this is as good a game to play as any other. But the main reason why so many good minds are still duped and play the game—be it academic philosophy or academic science—is that there are no viable alternatives. (Now there are many more than there were 20 years ago.) Young minds are very impressionable and they just follow the path. It takes a strong and a rebellious mind to stray from the established path.

Well already in the early 1960s I found analytical philosophy arid and suffocating. I found an antidote to it in the philosophy of Karl Popper. Popper was teaching in London, at the London School of Economics, while I was finishing my dissertation for Oxford. In order to shake off the incestuous climate of Oxford, I moved to London and started attending Popper's seminar. This was my path to liberation. At first a limited liberation—from the virtuosity of analytical-linguisitic pirouettes of Oxford philosophy. And then gradually from the straight jacket of the entire Western tradition.

Actually, the importance of Popper should not be

underestimated. He is at present the greatest living philosopher. That may not be saying much, as the 20th century did not produce many philosophical giants. Popper was a revolution in philosophy. *He shattered the rational shell of Western philosophical tradition and this shell has not been mended ever since.* This shattering is not so obvious to many. So perhaps, we will need to say a few words about it. Popper was a great pupil of Einstein. Not in the literal sense but in the metaphorical one. Popper surmised that since Einstein proved that Newtonian physics is not absolute or permanent, there is no reason to assumed that any form of knowledge is permanent or absolute. Thus all knowledge, including scientific knowledge is conjectural and tentative. The consequences of this position are far reaching. One of them is unwillingness of philosophers and scientists to talk about truth and reality. For what is truth and what is reality if all knowledge is tentative and conjectural. In short, the rational bottom has fallen out of science and the whole Western tradition and we still don't know it.

Another consequence which was carried out by Thomas Kuhn is the idea of scientific paradigms. Kuhn's paradigms is a direct outgrowth of Popper's phisolophy. We nowadays all talk about paradigms and paradigm shifts. And some talk about Buddhism as a new paradigm for life and philosophy combined.

Now one of the explicit principles of Popper's methodology is *relentless criticism:* you always have to be critical towards all theories including your own: always submit them to scrutiny. This attitude allowed me first to liberate myself from the constraints of analytical philosophy; and in due time from the limitations of Popper's philosophy as well. Actually by embracing Popper's idea of relentless criticism, I was subconsciously embracing one of the main principles of Buddhism: *be forever aware:*

*forever watchful.* Popper's methodology represents a limited case of Buddha's watchfulness. The methodology of the critical path may very easily lead to the path of the awakened one.

Let me move now briskly through the stages which led me to Eco-Philosophy. I arrived in Los Angeles in 1964. And what a shock it was to find myself in the middle of endless freeways after the sleepy spires of Oxford! I first started teaching philosophy of science, within the Popperian framework. But soon I developed a new mode of reflection which I called Philosophy of Technology. It became very clear to me that what was shaping the destiny of civilisation was not science but technology. Fifteen years later the Philosophy of Technology became recognised as a new field of philosophy.

After having spent in California about four years I became aware that it was not technology that was shaping the destiny of civilisation but something else. This something else was *Flower Children.* They appeared in Los Angeles on Sunset Boulevard in 1968 with nothing of offer but a single flower which they were giving to passers by. This was a revolution. It was a re-enactment of the act of Cordelia in *King Lear.* She promises nothing to Lear. She comes empty handed. But what she offers is *Love.*

I lived, at the time, in West Hollywood, amidst the ferment and the hippie happenings. I listened to their arguments. It tried to understand. Curiously, it was during these informal sessions, full of sound and fury, when lots of nonsense was spoken (as well as some sense) that I started to form the backbone of new ideas.

The recurring question which was coming over and again was: where did we go wrong? The issue was not whether we have gone wrong, but *where* have we gone wrong. Quite innocently they would turn to me and ask: you are a philosopher, you tell us.

Of course from the standpoint of the rational tradition which originated me, we haven't gone wrong at all. The position of the rationalist was, and still is: if we have some problems, it is because we are not rational enough. If we become more rational, then we shall solve these problems. The unstated premise was: if present rationality does not work well enough, we have to bring more rationality.

What was beautiful in those young people in the late 1960s was that they saw through the whole thing—the whole fallacy of Western rationality, the whole hypocrisy of Western life style, the whole dance that was supposed to bring us to the gates of the Paradise and instead it was bringing us to the gates of unmitigated existential hell of individual and social alienation.

While thinking through these problems I came to the conclusions, already in the late sixties, that the problem lay in the very foundation of Modern Western philosophy, in the visions and systems of Bacon, Galileo, and Descartes. As the result of my critical thinking over the foundation of Western philosophy, I published a number of essays which finally appeared as the book entitled: *Technology and Human Destiny,* published by the University of Madras Press in 1983.

Well, you cannot drag the dead horse of positivism forever. We are very good in the West at criticising. Give me any idea of system and I can criticise it ad infinitium, or ad nausea. In the mid-1970s, it became obvious that the instant revolution which the hippies and the Ecology Movement were proclaiming was not forthcoming. Instead the backlash of the conservative forces started to occur. Yet, the agenda was unfinished. The limited, onesided Western philosophies still exerted an enormous influence through academic institutions and structures which were

**Diagram I**
**Mandala of Eco-Philosophy**

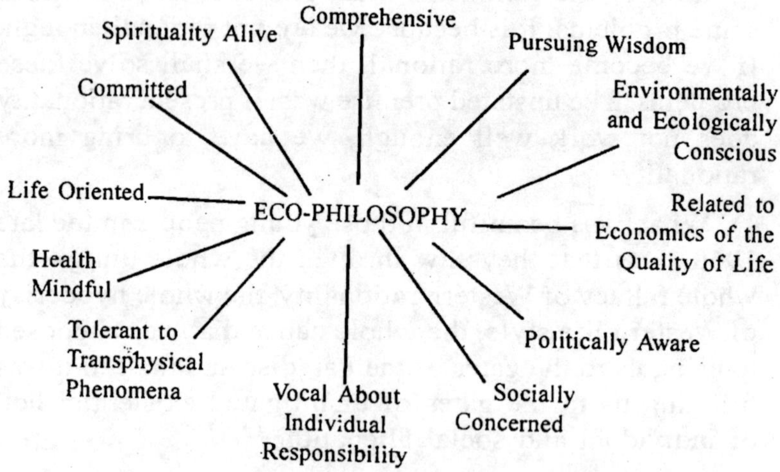


**Diagram II**
**Mandala of Present Philosophy**

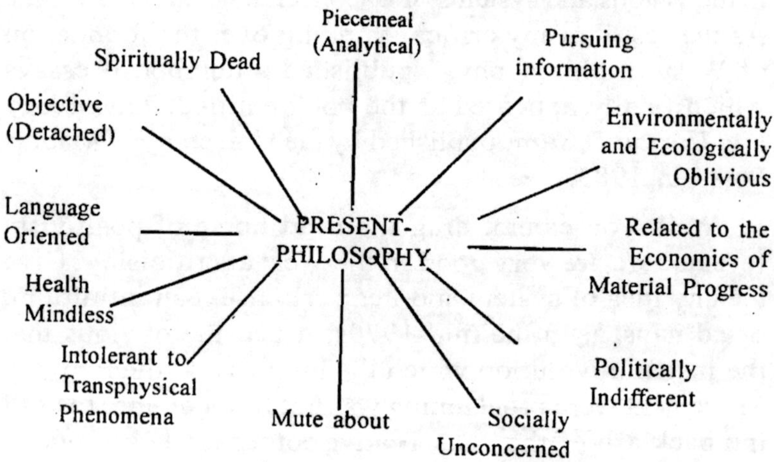

based on them. Let us be aware that although positivism may be dead as a specific philosophical doctrine, it is very much alive as far as scientific methodologies are concerned. The whole Western system of education is still based on the atomistic, analytical, piecemeal thinking which in virtue of its detached, objective, uncaring mode is geared to exploitation, not to healing.

So in late 1970, I started to search for an alternative. Given that the assumptions of Western mind and Western science, as conceived in the 17th century are inadequate, given the fact that the wholistic interdependent thinking which the Ecology Movement suggested is desirable and point in the right direction—what kind of philosophy, what kind of a new system should we construct which would be able to confront most of our urgent problems: of individual alienation and social disintegration; of the environmental crisis and the search for meaning? What kind of philosophy do we need which connects life with thought? What kind of philosophy may help us to live? The result of my grappling with these questions was the book *Eco-Philosophy, Designing New Tactics for Living,* published in 1981. Perhaps, the kernel of this philosophy is contained in the two mindless, contrasting Eco-Philosophy with the dominant analytical philosophy. (See diagrams I and II)

Hidden in the mandala of Eco-Philosophy is the Buddhist Eightfold path, of which I became aware only after I published the book. Let us remind ourselves of the formulation of the Eightfold path.

**The Four Noble Truths**

1. Know the true nature of suffering
2. Know the true cause of suffering
3. Know the true nature of disappearance of suffering

4. Know the right method leading to the disappearance of suffering

**The Eightfold Path**

(being the articulation of the 4th truth)

1. Right views (beliefs, assumptions)
2. Right thought (attitude of mind)
3. Right speech
4. Right conduct (action)
5. Right livelihood
6. Right effort
7. Right mind control
8. Right mediation

The question which has preoccupied me in recent years with increasing urgency has been: *What is the eightfold path appropriate for our times?* Subconsciously my entire opus on Eco-Philosophy was constructed to answer this question. Now, if we are to live our lives with the modicum of meaning, then some fundamental questions must be entertained. Among them are:

What is Appropriate Livelihood?

What are appropriate forms of thinking that genuinely support appropriate livelihood?

What are Appropriate values that follow from Appropriate livelihood, and support it?

What is the appropriate world view, an appropriate cosmology which provides space and Framework for Appropriate thinking, values, and livelihood?

This last question I consider very important, for unless we have the right foundations, out of which right values and action will follow, we will be trapped in isolated boxes that vitiate our quest for meaning and our quest for right livelihood. Thus, ecological cosmology which is the foundation of Eco-Philosophy is also the foundation of right livelihood; as well as the foundation of right thinking which now becomes ecological thinking or reverential thinking. Eco-Cosmology is also the foundation of right values which now become ecological values based on the idea of reverence for life and for all forms of creation.

While meditating over the foundations of a right philosophy, I asked myself the question: Why has Buddhism survived better than Christianity? The answers that gradually came to me were:

One, because it was based on a deeper and more encompassing vision than Christianity. Buddhism has been inclusive rather than exclusive, while both Judaism and Christianity have traditionally been exclusive.

Secondly, Buddhism survived better because it was consmos-centred *vis à vis* Christianity which was individual God-centred.

Thirdly, Buddhism survived better because it conceived of responsibility in global and universal terms, in terms of universal Karma; while both Judaism and Christianity have been more parochial, relating, our Karma too closely to our individual sins.

Fourth, Buddhism has been more attuned to nature and cosmos (than Christianity) as it has conceived of man as a micro-cosmos of the entire cosmos, and an inherent aspect of it.

Fifthly, Buddhism has been able to merge right meditation and yoga with right livelihood, while in the West our meditation and thinking are so often apart from our livi-

In short, Buddhism has been able to construct and maintain one integral system (with meditation and spiritual practice at the heart of it) while neither Judaism nor Christianity have been able to construct such an integral system.

What about the Christian prayer, were may ask? It has lost its power. We don't know how to pray. We only know how to ask God to give us more of this and more of that. A far cry from the universal compassion which the Buddhist creed advocates and the Buddhist mind attempts to obtain.

The universal compassion or the great compassion may be the single most important attribute that may save humanity. We need vehicles and philosophies that will teach us the great compassion or which will keep us to relearn it. Eco-Philosophy offers itself as one such vehicle.

# PART II
# ETHICS OF ENVIRONMENT

# 7

# NATURE AND SCOPE OF ENVIRONMENTAL ETHICS

Environmental ethics consists of the study of normative issues and principles relating to human interactions with the natural environment and to their context and consequences. It forms a crucial area of applied ethics—crucial for the guidance of individuals, corporations, and governments in determining the principles affecting their policies, lifestyles, and actions cross the entire range of environmental issues. It is equally crucial for the appraisal of such actions, lifestyles, and policies.

Sometimes the phrase "environmental ethics" is used, contrary to the previous definition, to refer simply to the quality of people's behaviour insofar as it relates to the natural environment. However, if this were held to be the central meaning of "environmental ethics," the implication would be that ethics consists of (ethically significant) behaviour, that ethical principles are invariably manifested in people's conduct, and perhaps also that no critical study of normative ethics is possible. While the usage just mentioned has to be acknowledged, it is important that there are also other uses of the phrase "environmental ethics" which concern not just such behaviour, but also the normative principles applicable to it and their critical study. This critical study is standardly known as "environmental ethics," the subject of this overview.

Environmental ethics is sometimes defined instead as the kind of approach to environmental issues which finds independent value to be located not only in the interests of sentient creatures—creatures which can feel pleasure or pain—but also in natural living creatures (or in their good) in general, or in the natural world in general. While many influential philosophies are committed to this approach, many others are not, and base their justifications on the interests so sentient creatures only. Since, the latter kind of approach is the approach of many environmentalists, and undeniably offers both specifications of environmental problems and solutions to them, it is advisable not to adopt a definition of "environmental ethics" which treats this approach (let alone more limited approaches) as lying outside environmental ethics. In this way, the debate about the location of independent value can continue to take place within environmental ethics, and the boundaries of the subject or discipline of environmental ethics avoid being treated as themselves a battleground about values. Also environmental ethics can be allowed to be a neighbour of, e.g., health-care ethics or business ethics, concerned about a different sphere, but not a rival discipline with essentially different values of its own.

While environmental ethics is distinguished by its sphere rather than its values, its sphere has made it much more alert than ethics has traditionally been to the interests both of future generations and of nonhuman creatures. Where the interests of coming centuries used to be neglected in general theories of normative ethics, such neglect has now become unsustainable, and this is at least partly due to several pioneering works in environmental ethics. Meanwhile the anthropocentrism of traditional ethics has widely been modified so that at least sentient interests are taken into account, as a result, not least, of the work of philosophers such as Peter Singer and Tom Regan. Thus environmental ethics tends to stress the interests both of future generations and of nonhumans. Meanwhile anthropocentric philosophers are increasingly aware that they hold a minority position in normative ethics, which stands in need of defense.

## THE SCOPE OF ENVIRONMENTAL ETHICS

The scope of environmental ethics is as extensive as its sphere: the realm of actions, policies, and lifestyles which impact on the natural environment, together with their contexts and their consequences, and the principles and attitudes which underlie these actions, lifestyles, and policies. The spatial extent of its scope is at least as extensive as the biosphere of planet Earth, if not larger, for space probes and their related debris result from human interactions with nature, and have extended the realm of environmental ethics to distant zones of the solar system. On the other hand, abiotic (nonliving) nature is not usually regarded as a principal concern of environmental eithics, at least not for those theories of value which find no value where there is no life, for in lifeless realms there is nothing which can be harmed or benefited. However, philosophers who locate value either in abiotic nature or in nature in its entirely are consistent in including in environmental ethics human impacts on distant starts and galaxies, however long delayed they may be.

The temporal scope of environmental ethics extends for as long as human action can exercise any kind of impact, and as long as something of value remains on which significant impacts can be made. While this may sound like a purely theoretical point, nuclear energy generation and experimentation is already bringing into being radioactive substances with half-lives of millions of years, with predictable harmful effects on almost any living creature which comes into contact with them. Hence, the temporal scope of environmental ethics turns out probably to transcend the future existence of humanity, in that its reasoning concerns the long-delayed impacts of current and of projected policies; only on an anthropocentric theory could nothing of value be affected for good or ill after humanity becomes extinct. This means that normative ethics, which used to be preoccupied with such localised problems as the returning of misdirected postal packages, cannot now avoid comprehending vast tracts of space and time if it is to be seriously applied to current practice.

Environmental ethics also studies the past in order to discover the traditions which often underlie current values, and which often turn out to supply limits to possible change in ethical attitudes or resources for such change. The rediscovery of past ways of life, or the discovery of historical saints or heroes, whether in our own culture or in alien cultures, can facilitate present imitation; the discovery of factors which have suppressed desirable patterns of living can generate powerful movements of present resistance. While environmental ethicists need to beware the invention of fictitious pasts, there remains an important ethical component to environmental history.

Fictional utopias, however, should not be considered irrelevant. Depictions of possible worlds or of possible communities living more harmoniously with nature ("ecotopias") comprise a further part of the scope of environmental ethics. They have to be scrutinised, however, for their consistency, and for adverse side effect which their nonfictional adoption might generate; the work of ethicists is seriously incomplete if such scrutiny is neglected.

Yet, this very exercise shows how environmental ethics is not confined to the actual past or future; counter factual judgements, ones relating to thought experiments and to the possible rather than to the facts of history or of science, are frequently indispensable when normative theories need to be appraised.

## ORIGINS OF MODERN ENVIRONMENTAL ETHICS

The widespread acceptance in the late 19th century of Charles Darwin's theory of evolution by natural selection made possible a recognition of the interdependence of living species, and, relatedly, of the extent of the unintended side effects of human action. The perceived importance of preserving tracts of wild nature led to the setting up of National Parks such as Yellowstone (in Wyoming) and Yosemite (in California).

But proposals for an extension of ethics to cover all the species of the living systems of the Earth first emerged in the 20th century, in Aldo Leopold's *A Sand County Almanac.* In Leopold's Land Ethics, the land is the community of the interdependent species of the planet, plus the other components of their ecosystems. It was Leopold's proposal that "a thing is right when it tends to promote the integrity, stability and beauty of the biotic community. It is wrong when it tends otherwise".

Environmental concern became much more widespread during the1960s, with increasing alarmig being voiced about nuclear fallout, population growth and its impact, and pesticides, as in Rachel, Carson's *Silent Spring* (1962). In 1973, the Norwegian philosopher Arne Naess classified ecology movements by their depth in a celebrated article, and related issues were presented to a World Congress of Philosophy by the New Zealand philosopher Richard Routley, based at Australian National University, Canberra.

The distinctive nature of such a new ethic was investigated in a ground breaking paper by Holmes Rolston of Colorado State University. By this time, an environmental philosophy conference had already been convened by Willion T. Blackstone at Athens, Georgia, and environmental ethics courses were already being taught at Universities, both in Wisconsin and in the University of Wales at Cardiff.

The first major journal in the subject, *Environmental Ethics,* was founded at the University of New Mexico in 1979, with Eugene C. Hargrove as Editor-in-Chief; Rolston serves as Associate Editor. By now there are a number of other journals in the field, such as *Environmental Values* based at the University of Lancaster, and work in this field is increasingly found in mainsteam journals of philosophy. But the subject owes much to Hargrove's dedication, resilience, and breadth of editorial vision. Having moved in 1981 to Athens, Georgia, and having weathered a controversy about his own tenure there, Hargrove later moved to the University of North Texas, where the journal continues to be based. Meanwhile in 1990 Rolston,by then the author of the leading monograph in the field founded the International Society for Environmental Ethics, of which he became President and Newsletter Editor, with Laura Westra (University of Windsor, Ontario) as Secretary. The Society has organised sessions at many conferences all over the world, and has a worldwide array of representatives. Its newsletter continues to carry a remarkable wealth of news and reviews.

Thus, the discipline of environmental ethics is by now institutionally entrenched in many parts of the developed world, and is increasingly chosen by philosophy researchers as the subject of their dissertations. Third World scholars too have played an active part in its development. The late Henry Odera Oruka was the founding director of an Ecophilosophy Centre at Nairobi, Kenya, and organised in Nairobi in 1991 a World Conference of Philosophy on the themes of environment, development, and their philosophical bearing. (It was Odera who drew my own attention to environmental philosophy in 1975.) A further example is supplied by the International Conference on Development, Ethics and the Environment, organised in Kuala Lumpur, Malaysia, in 1995 by Azizan Baharuddin of the University of Malaya and of the independent Institute for Policy Research. Third World environmental philosophers frequently stress the importance of blending environmentalism with the need for economic and social development to remedy the blights of poverty and injustice. Priorities between these diverse values continue to be debated.

It will already by clear that not all practitioners of environmental ethics argue from interests independent of human interests. Nevertheless, most environmental ethicists do not confine moral standing to human beings and organisations, thereby distansing themselves from anthropocentrism (in the sense previously defined). Sometimes, this is argued on the basis that many nonhuman animals are sentient, and that their interest in not being made to suffer has to be regarded as morally relevant; indeed discrimination on the sole basis of species membership has been labelled "speciesism" and compared to other kinds of discrimination such as sexism and racism. Others appeal beyond sentience to the capacity of all living organisms to develop and flourish in the manner of their own kind, or again claim that rights belong to species, or even to ecosystems.

Anthropocentrists respond that human beings cannot help appropriating resources from the natural world, and thus prioritising their own interests, and that, as agents, they inevitably operate from within a human-centred perspective. However, these claims are consistent with a nonanthropocentric stance in ethics; an ineradicably human-centred cognitive perspective is compatible with recognising ethical considerations independent of human interests, whether motivated by identification, by sympathy, or by administration of nature's otherness. Others claim that the underlying ground for identification with nature is the good of the person whose understanding is thus enlarged and humanised, but his stance has been accused of narcissism, and also seems blind to the independent significance of, e.g., animal suffering.

With greater sophistication, others again call "anthropocentric" those judgements of intrinsic value for which no objectivity is claimed, labelling them in this way simply in virtue of their being made by human beings, and designate the kind of value thus ascribed "weak anthropocentric intrinsic value"; the scope of such value is, they hold, by no means confined to human interests and their objects. But since, this is a different sense of "anthropocentric" from the usual sense, conveying a metaethical rather than a normative classification of value, and rather than support for anthropocentrism in the traditional, unsophisticated sense, clarity is better served by using some other term (such as "anthropogenic") for judgements and theories of this nature, and retaining "anthropocentric" for its traditional role.

Among those who reject anthropocentrism, another debate is between those who regard all individual creatures with moral standing as having equal moral significance and those who recognise deferring degrees of moral significance, related usually to deferring capacities (such as sentience and intelligence). The egalitarian camp regards belief in degrees of significance as discriminatory and arbitrary; the other camp responds that environmental justice demands that priorities be observed when clashes of interest occur. While the kind of biospheric egalitarianism which makes each organism count equally may be a theoretically consistent position, it suffers from the handicap of making life unlivable in practice.

Another debate arises between (on the one hand) campaigners for animal welfare and (on the other) campaigners for the conservation of species, ecosystems, or wilderness. This debate often corresponds to a radical difference of values, with animal welfarists arguing from the importance of the well-being of individual sentient animals, and conservationists, when not appealing to human interests, sometimes arguing that species or eco-systems have some kind of independent, holistic value. While these camps are allies for many practical purposes (as over preserving the last members of an endangered species and their habitats), they often diverge, as over the culling of deer in Scotland, of seals in Canada, and of elephants in southern Africa.

Debates on this and other issues among environmental ethicists often turn on "intrinsic value" (the kind of value which is neither instrumental nor in any other way derivative, but depends entirely on the nature of its bearer) and on its extension or location. Both animal welfarists and many conservationists reject an anthropocentric view of its location. But while the former locate it in individual well-being, some of the latter locate it in diversity, in wildness, in independence from human impacts, or simply in nature (understood as the realm independendent of human agency), and sometimes suggest that pain and suffering are intrinsically neutral (rather than bad), besides being instrumentally good in so far as they contribute to the maintenance of esosystem stability. However, the value of diversity seems to depend on its (by now controversial) contribution to stability; the criterion of wildness would deprive all domestic animals of intrinsic value; and the independence criterion elides the distinction between abiotic nature and living creatures. And if pain does not count even in general as an independent reason against what causes it, it is difficult to see what else counts. This holds true irrespective of whether the pain is experienced by a human being or by a nonhuman animal.

Yet, this is not to suggest that welfarists have nothing to learn from the conservationist camp. Conservationists, for example, can explain the importance of prediction tc the existence of both predators and prey, and thus why human beings should seldom if ever intervene to prevent it. Prediction, parasitism, and such-like apparent evils turn out to carry a positive, albeit derivative, value for the species and the systems involved. Further, if human society is to attain sustainability, and avoid undermining the interests of future generations (whether of human beings or of nonhuman creatures), ways must be found of not undermining the ongoing operation of those natural systems on which human life depends. Unless welfarists grasp the networks of inter dependence pervading the natural worked, their contribution to environmental ethics is fatally flawed. This, however, need not make their principled objections to factory farming, or to hunting as a sport, any less valid.

A further debate concerns whether environmental philosophy should concern itself with values or with ethics at all. In face of trenchant criticisms from Richard Routly concerning deep ecology, with its belief in the value of nature as a whole, and in the self-realisation which this belief makes possible, Warwick Fox claimed in its defense that talk of value on the part of deep ecologists such as Arne Naess and George Sessions should not be taken literally, that advocacy of beliefs about value and ethics was futile, and that the real message of deep ecology concerned the identification of the self with the greater Self of nature. Once self-realisation of this kind had been achieved, agents would in any case be motivated to defend nature, and no purpose would be served by ethical talk or talk of values. Responses have mentioned the self-undermining character of identification with nature as a whole, the lack of guidance available once values are discarded, and the need for interpersonal reasons implicit in values if actions are to respond proportionately to current problems. Deep ecology also seems to underply nature's otherness, an equally important source of environmental concern as identification and empathy.

Besides enriching value theory and normative ethics through its stress on future generations and on nonhuman creatures, environmental ethics has had to examine the status of its own claims, and has thus breatheed new life into the discipline of metaethics. Objectivists hold that claims about value, rightness, and obligation admit or truth, and comprise interpersonal reasons for action as opposed to comprising expressions of taste or prescriptions. Objectivists usually also hold that knowledge is sometimes possible in these areas. Subjectivists, by contrast, maintain that value is always value for someone or some group, or in some valuational framework, and that ultimately there are no merits by which such valuings or frameworks can be compared. Such views appear to deprive claims about value or about obligation from supplying reasons for action capable of being treated seriously.

Recently, the claims have been made that developments in quantum physics have made the distinction between subjects and objects untenable, and that the discourse of value must be understood as relative; the very language of objective value is claimed to belong to an obsolete modernist paradigm. It has been replied that quantum physics itself presupposes the language of objectivity, and does not require a subjectivist interpretation. Another discipline concerned with value, economics, conventionally treats value as a function of preferences; the relations between economic value and intrinsic value currently form a growth point for environmental ethics. An important distinction which has been made in this connection distinguishes our valuations as consumers (typically functions of our preferences) and our valuations as citizens (usually unrelated to preferences or to willingness to pay).

Related debates concern whether any single theory of value or obligation should be accepted, or whether working with a plurality of theories might be preferable to this quest for closure, and also whether the very concept of truth should be relativised to kinds of discourse. Such debates continue to figure in *Environmental Ethics* and in mainstream philosophy journals too, such as *The Monist*.

Other relevant debates concern the nature of the environment and of nature itself. These apparently abstruse questions turn out to have a crucial bearing on practice, as different answers to them prompt deferent policies. Thus if (as is often held) the environment is always *someone's* environment, then environments are always local, and much depends on the perspectives of the local human inhabitants. Principles concerning the global environment thus depend on some less localised concept of environmental and on the existence of a globally shared environment on which local environments depend.

By contrast with the concept of environment, the concept of nature (traditionally contrasted with convention or with culture) might seem to demand an objectivist understanding. Yet, the perspectives of different cultures about what is natural are so various that for many theorists nature and the natural are to be regarded as constructions. This view has important implications both for the value of nature (understood as something utterly independent of human agency) and for the project of preserving natural wilderness, and could potentially undermine the very ideal of identification with nature. In some ways, this debate overleaps that between realism and idealism in the theory of knowledge, but even if realism were finally vindicated, that would not undermine all claims about the constructed character of many ideas of nature. However, since constructivists believe that it is embodied people who construct such ideas, there is a significant inbuilt self-limitation to their claims about natural entities as constructions.

Feminist writers have often compared traditional attitudes to nature and to women, and complained of the oppression of both. Sometimes, it is also held that through their roles as lifegivers (childbearing) and nuturers. women are closer to nature than men, and that current problems will remain insoluble unless related feminine qualities (such as caring) are allowed to mold social life and policies. Debates about such ecofeminism and its bearing on the concept of nature continue in journals such as *Environmental Ethics.* The patriarchal nature of much of the inherited discourse about nature is undeniable, and there are important parallels between the past exploitation of women and that of nonhuman creatures. It is unclear, however, whether gender-related qualities as such exist, or whether the claim that there are such qualities may not be a further instance of gender stereotyping. What is more cogent is that alliances can be formed between the different movements against oppression, including, importantly, feminism.

## RELIGIONS AND ENVIRONMENTAL ETHICS

Recently, Max Oelschlaeger has argued that the best hope for the environmental movement in the United States is to enlist the support of the churches. He cites widespread support for the preservation and conservation of nature as God's world, both among believers and (more surpringly) among nonbelievers. These findings also generate surprise in view of the controversy concerning the supposed religious origins of the environmental crisis.

Nearly, three decades ago, Lynn White, Jr., argued that the roots of our ecological crisis lay in the Jewish and Christian doctrine of creation, or rather in the activists variant of it distinctive of the Western churches, as opposed to the Eastern Orthodox churches. The impact of this doctrine was delayed until post-Renaissance science became allied with technology in the 19th century, but its anthropocentric and despotic tendencies were already present in the early middle Ages, Other historians have questioned White's account of medieval history, while theologians and historians of ideas have disputed White's interpretation of the Bible. By now, it is agreed that theistic religions often regard humanity as steward of creation, answerable to God for its care, and that this stance has ancient origins. Disagreement concerns the extent to which it was or was not a characteristic view until the current century, and the extent to which those who adhered to it were or were not free of exploitative attitudes themselves.

While Christian attitudes to nature in previous centuries have been too various to fit White's rather simplistic interpretation, the debate about White's theory is often seen as a distraction from the theory and practice of problem solving in the present, in the United States, most of the churches are nowadays strongly environmentalist and strong enough to exercise considerable political influence, and even philosophers and ethicists committed to the independence of ethics from religion may need to take all this into account, if only on pragmatic grounds. (Whether this kind of environmentalism encompasses concern for the environment of the Third World or for the global environment remains to be seen.)

In most European countries, religion is not an influential as it is in North America. But Oelschlaeger's theory still has an international relevance, if suitably adjusted, for there are currently influential voices in Islam (as at the 1995 Kuala Lumpur Conference) representing a stewardship interpretation of the faith as historically central and as ethically mandatory. Nontheistic religions also have their environmental relevance, but it is the activists theistic religions which could transform the prospects for conserving the global environment.

## CAUSES OF ENVIRONMENTAL PROBLEMS

The identity of environmental problems (problems resulting from human interactions with nature) depends on values; for example, nonanthropocentric theories are much more concerned with extinctions of species and habitats than anthropocentric theories. Nevertheless, problems such as pollution, resource depletion, loss of cultivable land, and loss of wilderness are often assigned common causes. For example, the theory that religious attitudes are to blame for these problems has often been put forward, without securing agreement. More material causes have to be considered.

Population growth is often suggested as a principal cause, and many ecologists actually favour reductions of the human population and international sanctions to secure such reductions. Yet, environmental problems are more concentrated in areas of intense industrial activity than in areas of population concentration, and average impacts on the environment are often many times greater in the former areas than in the latter. While some problems can be correlated with population growth, this kind of growth in turn is often driven by poverty, and these problems are unlikely to be resolved until poverty itself is tackled. Meanwhile the causes of problems in industrial areas cannot be set down to population growth; for such problems, the proposed causes include affluence and also modern technology, and proposed solutions accordingly sometimes commend an ethic of simpler lifestyles reliant on technology of a simpler and less consumptive kind.

Yet economic forces, rather than individual consumption levels, are likely to drive the polluting processes, and this has led to advocacy of limits to economic growth, and sometimes of abandonment of modern technology. But humanity is unlikely to be fed, let alone global problems to be solved, without the aid of modern technology, and so not all growth should be rejected. Nor would the withdrawal of individuals or regions into self-sufficiency contribute much to the solution of these problems. The problems have to be understood against the background of the current inequitable international economic order, and are unlikely to be solved unless this order is radically restructured. Ethical theories in which all this is neglected are likely to prove transitory.

## SUSTAINABILITY

Where environmentalists in the 1970s used to advocate limits to growth, their successors from the late 1980s have instead often been advocates of sustainable development. Sustainable development was the central theme of the Brundtland Report, where it was defined as "development that meets the needs of the present without compromising the ability of future generations to meet their own needs." Besides stressing that not all growth should be regarded as beneficial change (or development), the authors of this report stressed that development must be sustainable, and thus inaugurate or fortify economic and social processes capable of being continued indefinitely, without undermining either themselves or the ecosystems on which both nonhuman creatures and human economic systems depend. The United Nations Conference on Environment and Developmental Rio de Janeiro (1992) took some promising steps toward concerning this notion into reality.

Advocates of sustainable development, however, encounter problems of definition, and also opposition from diverse directions. When they attempt to specify what precisely is to be sustained, one account,which makes this the economic value of natural resources, would allow the elimination of species and wild habitats wherever this would result in (say) enhanced technological options for humanity, while another, which would forbid all such changes, would prevent almost all development in regions (for example) of reinforest, at the cost of failure to satisfy basic human needs. Sustainable development, insofar as it diverges from traditional economics, needs to blend the criteria of satisfying current human needs, allowing equivalent options for satisfying future needs, and preserving the bearers of intrinsic value (nonhuman creatures included).

Radical environmentalist would still criticise the concept and the pursuit of sustainable development for permitting "business as usual." and for being too easily appropriated into the vocabulary of capitalist enterprises. Certainly, conventional politicians and representatives of business often pay lip service to "sustainable development," meaning thereby nothing more than sustained economic growth. The radical critics sometimes urge an ethic which would reject development and/or industrialism altogether, and focus on stabilising economic processes. Meanwhile other critics object to potential departures from cost effectiveness, and claim that sustainability is not always a virtue. However, both sustainability and development embody values which, as their supporters are aware, should not be discarded lightly. Accordignly, work continues on both the theory and the practice of sustainable development at international, national, and local levels.

## SOME FRAGMENTARY SOLUTIONS

Ethics is concerned not only with ideal solutions, such as a radical restructuring of the world economy would involve, but also with what should be done while things remain largely as they are, by agents with limited powers and a limited scope for action. Despite their greater powers, governments and corporations are often in this position, rather as individuals are. The approaches which follow are suited to agents and agencies in situations of these kinds.

In view of the danger that environmental impacts will cross critical thresholds or prove irreversible, and in the light of evidence that environmental risks are often underestimated, a principle that has increasingly been accepted by European governments in recent years is that action such as regulation may justifiably be taken to avert environmental harm in advance of the availability of scientific evidence confirming that harm. This principle is sometimes supplemented by the requirement that the best available technology be used; its supporters, however, reject the inclusion of the qualification that this technology should not incorporate excessive costs. This Precautionary Principle is not a basic principle, is all too liable to be given an anthropocentric interpretation or a pro-western bias, and needs to be applied in conjunction with principles both of sustainability and of justice; yet its adoption involves little more than common prudence, and its earlier implementation could have curtailed acid rain, ozone depletion, and even global warming.

In this approach to environmental decision making, all the entities liable to be affected by a development or policy are identified, including future people and non-human creatures both present and future, and the probability and extent of the impacts are calculated (as far as quantitative calculation possible) and then aggregated, like interests being treated alike. Factors which can not be quantified are identified alongside those which can be when overall impacts are reported. This method is one tool among others for decision makers, who are also encouraged to consider giving additional weight to impacts affecting the poor and the vulnerable. In requiring the inclusion of nonhuman interests, this approach diverges from conventional economics; in this connection, human proxies would estimate impacts on the creatures concerned. A further divergence would inolve discounting future interests only where special reason (such as inflation) make this appropriate. This approach is presented in *Values, Conflict and the Environment,* its adoption would strengthen the quest for sustainability. The danger that monetary costings would be used too widely could be countered by frequent use of the nonquantitative category of impacts.

Global warming is now recognised to be caused by human activity, through the emission of greenhouse gases such as carbon dioxide and methane, and to be likely to raise the level of the oceans through the melting of polar ice caps to an extent which threatens the continued existence of islands such as the Maldives, and of low-lying areas such as much of Bengal, plus all their flora and fauna, human beings included. Even before scientific consensus was reached, applications of both the Precautionary Principle and of comprehensive weighing implied that constraints on carbon emissions were ethically obligatory, despite the interest to the contrary of energy-producing countries and corporations. Developing countries, however, cannot justly be expected to curtail carbon emissions until they are able to satisfy the basic needs of their citizens. Emission quotas must therefore be recognised and observed by the industrialised countries, whether through an aggrement or treaty, or (if this cannot be negotiated) through informal policy decisions. Negotiations, originating from the Rio Conference, are due to be concluded in Kyoto in 1997. Ethical arguments, such as those put forward by the Association of Small Island States (AOSIS), have played a significant role in persuading the representatives of both developing and industrial countries of the validity of these principles.

While an agreement on carbon emissions will not make the Earth an ecological paradise, rudimentary steps such as this one are ethically indispensable, and could supply a paradigm for the further measures which the global citizens of the future will need to take.

## REFERENCES

Attfield, R. (1992), *The Ethics of Environmental Concern,* 2nd ed. University of Georgia Press, Athens, GA.

Attfield, R. (1994), *Environmental Philosophy, Principles and Prospects.* Avebury, Aldershot/Ashgate, Brookfield, VT.

Attfield, R. (1995), *Value, Obligation and Meta-Ethics,* Rodopi, Atlanta, GA/Amsterdam.

Attfield, R., and Dell. K. (Eds.) (1996), *Values, Conflict and the Environment,* 2nd ed Avebury, Aldershot.

Evernden, N. (1992), *The Social Creation of Nature*, Johns, Hopkins University Press, Baltimore/London.

Fox, W. (1990), *Toward a Transpersonal Ecology: Developing New Foundations for Environmentalism,* Shambala, Boston/ London.

Hargrove, E.C. (1992), Weak Anthropocentric Intrinsic Value, *Monist,*, 75, 183-207.

Oelschlaeger, M. (1994), *Caring for Creation: An Ecumenical*

*Approach to the Environmental Crisis,* Yale University Press, New Haven, CT.

L'Neill, J. (1993) *Ecology, Policy and Politics,* Routledge, London/ New York.

Passmore, J. (1974), *Man's Responsibility for Nature,* Duckworth, London.

Regan, T. (1983), *The Case for Animal Rights,* Routledge and Kegan Pual, London.

Singer, P. (1976), *Animal Liberation, A New Ethic for Our Treatment of Animals,* Jonathan Cape, London.

Thompson, J. (1990) A refutation of environmental ethics, *Environmental Ethics,* 12(2), 147-160.

# 8

# ETHICS OF LIFE PROTECTION

People who have views of a moral kind about environmental issues are committed to an environmental ethic consisting of at-least one, but usually a number of principles. Consider environmentalists who say that the extinction of species as a result of human actions is a bad thing maybe even a bad thing no matter what the cause. This might be a basic principle in an environmental ethic. Without having explicitly represented it as such, an environmentalist might nevertheless be committed to the view that the extinction of species etc. is bad considered in itself, quite apart from any consequences it might have. Another possibility is that the principle is not itself basic but rests on a principle enjoining concern for human welfare, combined with the belief that humans are harmed by the extinction of species. Making the ethical commitment explicit is a first step in subjecting it to critical appraisal or justification. Justification is necessary if we are to adjudicate between the various competing environmental ethics we encounter. It is not enough that an environmental policy conform to the principles of some or other environmental ethic, it should conform to the correct, or best justified, one. We have two questions: how might an environmental ethic be fleshed out? How might putative environmental ethics be justified?

## HUMAN-CENTRED ETHICS

Some think that environmental policies should be evaluated solely on the basis of how they affect humans. This entails a human-centred environmental ethic. Although, the classical utilitarians include animal suffering in their ethical calculation, variant of utilitarianism, which enjoins us to maximise the surplus of human happiness over human unhapiness, is one example of a human-centred ethic. Taking this ethic seriously obliges us to calculate the varying effects of the Kakadu options on human happiness and unhappiness. We might discover that mining would reduce the ecological richness of the wetlands and that if this happened then some people would be made unhappy; for instance, some might be moved by the plight of individual animals. Some might be saddened by the loss of species, some including members of future generations, might miss out on the chance of particular recreational of aesthetic enjoyments, some might be adversely affected by resultant climatic changes, changes in flood patterns and so on, some might be psychologically harmed by the despoliation of areas to which they have a spiritual attachment. These negative effects would have to be subtracted from any increases in happiness which resulted from mining in Kakadu. A human-centred ethic could lead to substantial agreement with environmentalists about policy. This would depend on the facts about the effects on humans of changes to the natural environment. However, this decision would have been reached by considering the interests of humans alone. A helpful way of putting this is to say that this ethic treats only humans as morally considerable. Something is morally considerable if it enters into ethical evaluation in its own right, independently of its usefulness as a means to other ends. Consider the Pig-nosed Turtle. On the human-centred ethic just now described neither the species as a whole nor the individual members of it are morally considerable: it is only the happiness and unhappiness of humans which is morally considerable and this might or might not be affected by what happens to the turtles.

## ANIMAL-CENTRED ETHICS

There is a view of ethics which counts not only humans as morally considerable but non-human animals as well: it includes all animals in its scope. Many of the things which we do to the natural environment do affect non-humans adversely and this, it is suggested, must be taken into account. For example, if we thought that cyanide pollution in the South Alligator River would cause non-humans to suffer then this is a moral minus which must be taken into account independently of how things will be for humans. The example is not fanciful: consider the effect on non-humans caused by clear-cutting forests, damming river valleys, quarrying mountains, constructing pipelines and so on. An animal-centred eithic enjoins the moral consideration of individual animals not of species: what happens to species is only if indirect concern insofar as it affects individual animals.

While an animal-centred ethic counts all animals as morally considerable it does not necessarily rank them equally. A useful way of putting this is to say that some animal-centred ethics will accord different moral significance to animals of different kinds. One form this differentiation might take involves the arbitrary and many would say unjustified, discounting of the interests of non-humans simply because they are the interests of non-humans. Just how this will affect judgments about policy will depend on the degree of the discounting. It could be such as to make human interests always count for more than non-human interests no matter what the intensity or strength of the interests and no matter what the numbers involved.

It could be such as to allow stronger or more numerous non-human interests to prevail over weaker or fewer human interests. Avoiding arbitrariness seems to require that equal interests be treated equally. This would leave scope for differentiation, which might still be made on the basis of interests which not all animals have. For example, humans have a capacity for developing theoretical knowledge or for rational autonomous action, which are arguably not capacities of kangaroos. These capacities might underpin certain interests which, because they lack them, kangaroos could not have. Such additional interests might swing a decision in favour of humans and against kagaroos. This is particularly likely in, although not restricted to, cases in which their common interests are equally threatened or equally protected: the appeal to the additional, unshared interest acts as a tie-breaker. Imagine that some important medical breakthrough depended on confining either humans or kangaroos. Keeping kangaroos in a very large enclosure in order to study them may be morally preferable if it threatens no interests of theirs; if they are not treated cruelly, if they are fed, if they are able to behave according to their nature. Confining humans in the same way is not morally acceptable because of the additional interests of humans. This mode of differentiation treats equal interests equally regardless of species and it also allows that unshared interests leave room for degrees of moral significance.

## LIFE-CENTRED ETHICS

The class of living things includes more than humans and non-human animals: it includes plants, algae, single-celled organisms, perhaps viruses and, it is sometimes suggested, ecosystems and even the whole biosphere itself. The complexity of a life-centred ethic will depend on how the question 'What is living?' is answered. However, this question is settled it will make much of the idea of a self-regulating system which strives, not necessarily consciously, towards certain goals. Moreover, it is this feature which is typically supposed to confer moral considerability on living things. A life-centred ethic counts all living things as morally considerable, although not necessarily of equal moral significance. So, it might be better to save a Pig-nosed Turtle than a waratah shrub, even though both are morally considerable. The former, however, might be more morally significant because it is a more complex living thing. Here, complexity acts as an intensifier: if living, then the more complex, the more morally significant. To take a different kind of case, it morally might be preferable to save some plant rather than to save the Pig-nosed Turtle, because only that plant can fill its particular ecological niche, whereas the Pig-nosed Turtle fills a niche that perhaps could be filled by similar turtles of a different species. Here, the differentiation is based on a moral assessment of the respective consequences of the plant ceasing to exist and the Pig-nosed Turtle ceasing to exist, and not on internal characteristics of the living things themselves.

A life-centred ethic requires that in deciding how we should act we need to take account of the impact of our actions on every living thing affected by them. For example, if mining goes on in Kakadu, it will involve cutting down trees and destroying other plants: it will cause the death of some animals and impair, if not destroy, wetland ecosystems. These facts and others count against mining and collectively must be weighted against the good things that might result if mining does go ahead. Since, the good things would seem to include only material benefits for some humans, it would be difficult to do the evaluation sum in such a way that it sanctioned mining. This is not so say that it is never morally permissible to fell trees, to flatten dunes, to kill animals, to modify ecosystems and so on. Whether it is permissible depends on what the outcomes are and on differences in moral significance within the class of the morally considerable. A life-centred eithic, incidentally, might take a radical form: it might claim that not only are all living things morally considerable but also that they are of equal moral significance. This biotic egalitarianism, if it could be justified, would make it very difficult indeed to defend morally human interventions is the natural environment. It would allow only quantitative judgements; for example, that two living things count for more than one. Most proposed life-centred ethics allow for differential significance within the class of living things. Although, humans might not be counted the most significant. The preservation of the biosphere and of large ecosystems might be thought more significant than the preservation of large numbers of humans.

## RIGHTS FOR ROCKS?

The ethics so far considered each evaluate actions by considering consequences for individuals and adding them. What distinguishes these ethics are the kinds of individuals within their scope: moreover, the later ones include all individuals included by earlier ones in the list. It could be argued that we are drawn inexorably to a life-centred ethic: that there is no non-arbitrary way of stopping the drift from the ethic of narrowest scope to the ethic of widest scope. Why not take the argument another step and count non-living things too as morally considerable? There is no attempt here to attribute a mental life or a point of view to non-living things; that would be to enter into an entirely different dispute. The claim is that non-living things, which, like many living things, lack consciousness and which also lack even rudimentary biological organisation, are morally considerable. Call this ethic the 'everything ethic.'

Take rocks for example. Mining will involve smashing up rocks, disturbing geological structures, spoiling fossils and the like. Is there anything wrong in doing these things? Here, we must take care to forget for a moment the consequential damage which would be done to plants, animals and ecosystems; we must ask whether these things would be wrong considered in themselves. Another example might highlight the issue. Imagine a plan to test a missile by firing it at some distant and completely lifeless celestial body which will be thereby destroyed. Would this be wrong considered in itself? On the ethic which attributes 'rights' to rocks, so to speak, it would be. All things considered it might not be wrong, but according to this ethic that is a case which must be made. Like the life-centred ethic, this one can be fleshed out in a variety of ways. It may allow degrees of moral significance and attribute comparatively minimal moral significance to non-living things. It may mirror biological egalitarianism and deny that there are gradations of moral significance, or it may fall somewhere in between.

## ECOLOGICAL HOLISM

It was earlier said that any ethic which would guide us in our treatment of the natural environment is, in the most general sense, an environmental ethic. The term 'environmental ethic' sometimes has narrower uses. It is sometimes used to indicate an ethic which counts as morally considerable individuals other than humans and which provides some solid purchase for the moral demands of environmentalists. A life-centred ethic is an environmental ethic in this sense, an animal-centred ethic less clearly so. However, some reserve the term for a specific ethic, ecological holism, presumably because they think that only such an ethic provides morally satisfactory protection for the natural environment. Ecological holism counts two kinds of things as morally considerable: the biosphere as a whole and the large ecosystems which constitute it. Individual animals, including humans, as well as the plants, rocks, molecules etc. which constitute these large systems are not morally considerable: they matter only insofar as they contribute to the maintenance of the significant whole to which they belong. Why should we worry if some species is caused to become extinct? We should worry not because of what this implies for its individual members or even for the species itself but because the extinction runs counter to the goal of maintaining the biosphere or ecosystems.

It is a moot point whether ecological holism should be thought to differ structurally from the other ethics. They had as their focus individuals, and 'holism'might be thought to signal a different focus. However, it is possible to view the biosphere and ecosystems as individuals, albeit extremely complex ones. If so, the holism amounts to the view that individuals many have hitherto regarded as morally considerable are not. Note that, although the principles of ecological holism differ from those of the other ethics, this does not entails that it differs from all of them in its policy implications. The life-centred ethic and the everything ethic are likely to sanction similar environmental policies as a result of the nature of the mechanisms which maintain ecosystems and the biosphere. Also, it is possible to combine ecological holism with any of the other ethics described. If, for instance, it was combined with the animal-centred ethic we would be enjoined to consider the interests of animals and the goal of biospheric maintenance. Where these conflict, for example in some odd case where animals can only be saved by simplifying an ecosystem, then some kind of trade-off or balancing would be required.

## JUSTIFYING AN ENVIRONMENTAL ETHIC

It is not too difficult to appreciate what is compelling about the claim that humans are morally considerable. Most obviously they are considerable because they have interests which can be harmed or advanced. These interests are based on capacities which humans have; for example, the capacity to experience pain and pleasure, the capacity for rational choice, the capacity for free action. Less obviously, they are considerable because of properties or characteristics which they possess which do not give rise to interests, to things in which they themselves have a stake. For example, it might be argued that anything which has the property of being a complex living thing is, to that extent, intrinsically valuable, which is to say that there is a moral reason for preserving it for its own sake independently of whatever uses it serves.

What is compelling about a human-centred ethic pushed us toward an animal-centred ethic, possibly further. (This argument is developed by Lori Gruen in Article 30, Animals.) Consistency and the avoidance of arbitrary moral distinctions fuel the shift from a human-centred ethic to an animal-centred ethic. Also, in thinking about non-humans, we might notice new reasons for moral considerability; for example, non-humans might have aesthetic properties such as beauty, which we might think make them morally considerable. This, too, is a case where they are morally considerable not because they have interests but because they possess some property which gives time intrinsic value.

Do the reasons advanced in support of animal-centred ethics also support life-centred ethics? If plants (and ecosystems or the biosphere) can be said to have interests, such as an interest in continued existence, then perhaps they do. The concept of *interest* is often explained in terms of a thing having a good of its own which can be harmed or promoted. Some claim that plants have a good of their own; for example, that the good of a tree is promoted by sufficient nutrients for its continued flourishing and harmed when it is deprived of nutrients. A plant's good is determined by the kind of thing it is, by the type of biological organisation it exemplifies, by what it is for it to be a flourishing member of its kind. Plants have a good in this sense but this is obviously not enough to ground the claim that they have interests in any morally relevant sense. Plants do not have a point of view from which they experience the world. It doesn't matter to the tree that it whethers and dies from want of water; it would matter to a kangaroo. While plants have natural goals, they have no attitude to those goals and progress towards them is not something which they experience. Similar points can be made about the biosphere and about ecosystems. It is this difference which is thought by some to stop the drift, by providing a non-arbitrary cut-off, from an animal-centred to a life-centred ethic.

Even if it is denied that plants have interest, however, it does not flow that they are not morally considerable. Recall that there were reasons suggested, which did not have to do with interests, in virtue of which humans and non-humans are morally considerable. These concerned the property of being a complex living thing and the property of being beautiful. Plants can possess these properties and if animals are morally considerable in virtue of possessing them, then so too are plants. The key to defending thus a life-centred ethic is to establish that the properties appealed to are intrinsically valuable.

Is there anything that might be said in defence of a life-centred ethic which pushes us toward an everything ethic? The property of being a complex living thing cannot be exemplified by rocks etc., but a related property, namely that of being a complex system, can be exemplified by collections of non-living things exhibiting certain relationships with one another. If it is organisational complexity *per se* that makes something morally considerable, then some non-living things will be morally considerable; for example, the bodies which make up the solar system, patterns of weathering on a cliff and a snowflake. The relevance of this suggestion to the Kakadu case depends, among other things, on whether ecosystems count as living things. If they do not then they are non-living things which exhibit complexity and which, given the suggestion, are morally considerable. The fact that they are morally considerable would provide an ethical reason for opposing mining. Or again, we might judge that one reason we think that living things are morally considerable is because they exemplify beauty. In some cases beauty might be exemplified by a thing's more general, external features, as in the case of tigers, whales, orchids and proteas. In addition, beauty might be exemplified in the more specific detail of a thing's biological functioning. Now some non-living things such as boulders, dunes, lifeless moons and icebergs can be beautiful, so, if exemplification of beauty is a basis for attributing moral considerability to living things, then at least some non-living things are morally considerable. The claim that exemplifying beauty is a basis for moral considerability is contentious; however, it is strongly supported by some, for example Rolston, 1988. Those who oppose it typically urge that it is the appreciation of beauty rather than beauty itself which is morally significant.

So, one way in which the move from one ethic to the next is accomplished is by finding a determinant of moral considerability in that ethic and showing that a rigorous application of it leads us to the next kind of ethic. Another way is by showing that there are new morally relevant features which the more restrictive ethics unjustificably ignore. One such feature might be the property of being a natural object: that is, an object which is not the product of human technology and culture. Rocks are natural objects and so on this view it would be wrong, although perhaps not all things considered wrong, to destroy them. There are other candidate properties; for example, the property of exhibiting diversity of parts, the property of functional integration of parts, the property of exhibiting harmony and the property of being a self-regulating system. This last group of properties, if deemed determinants of moral considerability, move us in the direction of ecological holism or in the direction of a mixed ethic. This is because they are properties quintessentially exemplified by ecosystems and the biosphere. If we accept that they are determinants of moral considerability, then we are provided with a reason, in addition to any we might derive from the other ethics we have considered, for resisting policies which would lead to disruption of ecosystems.

How do we decide whether candidate determinants of moral considerability are in fact such? Consider naturalness and exhibiting diversity of parts. Imagine that a certain mine requires the destruction of a group of trees on a rocky outcrop and of the outcrop itself. Environmentalists protect that this involves an uncompensated loss of value. The mining company promises to reconstruct the outcrop from synthetic parts and to replace the trees with plastic models. This bit of artificial environment will be indistinguishable, except by laboratory analysis, from what was originally there. It will be exactly as appealing to look at, no animals will be harmed as a consequence and no ecosystem will be disrupted. Neither the human-centred ethic nor the animal-centred ethic provides space for an environmentalist rejoinder. The life-centred ethic does to the extent that it permits a complaint about the killing of living trees. However, this does not seem to some to be the only thing morally amiss with the mining company's proposal. Isn't it also morally suspect because it replaces that natural with the artificial? Imagine a modified case in which only a rock outcrop, devoid of life, is removed and later replaced with synthetic rock. Not even the life-centred ethic allows for a complaint about the morality of this. Some people think that even in the modified case the mining company does something to which a moral minus attaches. If this thought is persistent it provides support for a variant of an everything ethic which includes within its scope all natural items. It is difficult to be entirely sure about the source of the belief, if we have it, that naturalness is a determinant of moral considerability. It is possible that we think that there is something dubious about the artificial outcrop because we can not distance ourselves from the thought that it will be detectably different or from the belief that it will harm animal interests or that it will result in ecosystemic disruption. If these are the sources of our belief, then we have no basis for the view that naturalness is a determinant of moral considerability. There is another possibility to which we should be alert. Naturalness might be a conditional determinant: that is, it might require the presence of some other property, for instance complexity. So, it isn't natural items which are morally considerable but things which are both natural and complex.

Consider the property of having a diversity of parts. Is this a determinant of moral considerability? Here, we might compare an area which is covered with rainforest with an area which has been cleared of rainforest and is under cultivation. Which is more valuable in itself? Again, we must distance ourselves from certain thoughts; for example, the thought that clearing rainforests is contrary to long-term human interests, the thought that wild animals would have suffered during the clearing or the thought that aboriginal people were displaced. Having attempted this, many would say that the rainforest is of more value. Imagine, then, that only one of these areas could be saved from massive devastation. Many would say that, considering them just in themselves, the rainforest should be saved. Moreover, one reason that will be given is that the rainforest exhibits more diversity; it is constituted in a more complex, richer fashion. There are other reasons that might also be given: for example, that there are aesthetic properties possessed by the rainforest but not possessed by the cultivated area. Our preparedness, by the way, to attribute aesthetic properties, such as beauty, to the rainforest, may well depend on whether we have an understanding of it as an ecological system: knowing how the parts work in concert to maintain the whole might assist us in seeing it as a thing of beauty. Counting these kinds of reasons as reasons for avoiding environmental despoliation provides the basis for an environmental ethic which reaches beyond either a human or animal-centred one and possibly beyond a life-centred one as well.

Even if we accept, for example, that the ecosystems of Kakadu are morally considerable, how do we weigh this against human (or other) interests? A first step is to ask whether there are alternative ways of satisfying human interests. Often there are. Moreover, the modification of ecosystems is often contrary to long-term human interests. Sometimes, there will be cases of genuine conflict where the different moral considerations pull in different directions. Here, we must carefully enumerate the relevant moral considerations, ask ourselves how important they are and then make an all-things-considered judgment. There is no decisive calculus available to assist us in these judgements. It is not correct to say that humans should always come first nor is it correct to say that preserving an ecosystem is always more important than protecting any set of human interests. Nevertheless there will be cases, such as Kakadu, where the morally appropriate policy is clear enough.

## REFERENCES

Attfield, R., *The Ethics of Environmental Concern* (Oxford: Blackwell, 1983).

Baxter, W.F., *People or Penguins: the Case of Optimal Pollution* (New York: Columbia University Press, 1974).

Callicott, J.B., 'Elements of an Environmental Ethic: Moral Considerability and the Biotic Community', *Environmental Ethics,* I (1979), 71-81.

Goodpaster, K., 'On Being Morally Considerable, *Journal of Philosophy.* 75 (1978), 308-25.

Naess, A., 'Self-realisation in Mixed Communities of Humans, Bears, Sheep and Wolves'. *Inquiry,* 22 (1979), 231-42.

Norton, B., *Why Preserve Natural Variety?* (Princeton: Princeton University Press, 1988).

Rolston III, H.: *Environmental Ethics: Duties to and Values in the Natural World* (Philadelphia: Temple University Press. 1988).

Taylor, P., *Respect for Nature* (Princeton: Princeton University Press, 1986).

Van De Veer, D. and Pierce, C., eds., *People, Penguins and Plastic Trees: Basic Issues in Environmental Ethics* (Belmont, Cal.: Wadsworth, 1986).

# 9

# WESTERN VIEW OF ENVIRONMENTAL ETHICS

The environmental crisis in the United States arises out of the dilemmas of prosperity. In his study of America in the mid-nineteenth century, Alexis de Tocqueville observed: "Their ancestors gave them the love of equality and of freedom; but God Himself gave them the means of remaining equal and free by placing them upon a boundless continent."[1] Indeed, the natural abundance of this country has made it a "promised land" of biblical proportions—a cornucopia of riches which have, no doubt, helped shape the national character of its people. In the words of a recent president, "The breadth and variety and beauty of our land, the richness of our mines and soil and forests and water, the favourable nature of our climate—all of these natural factors have provided a setting in which the optimism, the ingenuity and drive of the American people thrive and grow and are rewarded."[2]

The price of plenty has also meant that it has been impossible for Americans to imagine a time when its "spacious skies" could be polluted, and its "alabaster cities" be reduced to asphalt jungles. Since, these problems were not anticipated, neither were prepared to prevent them. "Even as we are heirs to the products of our forefathers' genius, so our generation has also inherited the results of past carelessness."[3]

Actually, the problem is one of more than "carelessness." It is a defect which arises out of the strengths of American society—its science, democracy, individuality, and prosperity. Broadly, it is a flaw of our civilisation; and that ultimately boils down to how we, as Westerners, perceive ourselves in relation to nature.

This chapter addresses the challenge coming out of America's abuse of the environment: its failure to respect the delicate web we share with other forms of life, and its neglect of humane design in manufactured artifacts, most conspicuously in the area of urban planning. Our purpose is not to indulge in doom and gloom, but to go beyond mere survival to finding new ways of bringing quality into our lives and restoring the balance of nature.

We begin by isolating the main agent of environmental destruction. The picture is much more complicated, but if we are to narrow the field to a single cause it would have to be technology, or better stated, our handling of technology.

Modern technology is the special child of Western culture; that is to say, it required a particular ethos and ideology in which to take birth and flourish. Western philosophy set the stage with a built-in dualism between spirit and matter reaching back to Plato and earlier to the Orphic mysteries. This strain of classical thinking was passed on to the modern West through Renè Descartes (1596-1650), French philosopher and scientist. In his scientific investigations, he followed the methods laid down by Francis Bacon. His philosophy was almost completely dualistic. His view of the physical world was that of a machine, wholly different from the mind. Only through the intervention of God were the world and mind brought together. As the "father of modern philosophy," Descartes had an immense impact on the development of European thought, especially through the Cartesian philosophers, who chiefly studied the problem of the relation of body and soul, of mind and matter.

During the Enlightenment, the mechanistic interpretation of the world was symbolised by the workings of a watch. Watches came in diverse forms and shapes, but their movements were all predetermined. Once God, conceived as the heavenly Watchmaker, had created it, the machine was able run as determined. The sharp metaphysical line between humans and their environment was continued in the thinking of Berkeley, Kant, Hegel, and Sartre. None of these philosophers conceived of the subhuman world in terms that would entitle it to fundamental human concern.

Western religion has followed a similar path. The three major Western religions—Christianity, Judaism, and Islam—teach that the universe is the creation of God, who has given human beings a special place in that creation. The book of Genesis (1:28-30) describes the position of persons in relation to the rest of creation: and God said to them, "Be fruitful and multiply, and fill the earth and subdue it; and have dominion over the fish of the sea and over the birds of the air and over every living thing that moves upon the earth." And God said, "Behold, I have given you every plant yielding seed which is upon the face of all the earth, and every tree with seed in its fruit; you shall have them for food. And to every beast of the earth, and to every bird of the air, and to everything that creeps on the earth, everything that has the breath of life, I have given every green plant for food."

The story of Creation is followed by the story of the Fall. Later a wholesale purge of wickedness became necessary, and God sent the Flood, which exterminated not only the majority of the human population but innocent subhuman species as well. God then made a covenant with his servant Noah in which man's earlier dominance was reaffirmed, and he was also permitted the liberty of a carnivorous diet. Genesis 9:1-3 states: And God blessed Noah and his sons, and said to them, "Be fruitful and multiply, and fill the earth. The fear of you and the dread of you shall be upon every beast of the earth, and upon every bird of the air, upon everything that creeps on the ground and all the fish of the sea; into your hand they are delivered. Every moving thing that lives shall be food for you; and as I gave you the green plants, I give you everything.

The pattern of Western philosophy and religion to distinguish sharply between humans and nature, and thereby to objectify nature so that it is seen from the outside, was extended by Western science. The separation became the attitudinal basis for an ever-increasing knowledge of nature by science. In Western thought, nature was desacralised, and therefore considered an object fit to be studied by separating part from part, limb from limb. "This superlatively effective way of discovering solidly verifiable truths tends, precisely because it is sharply focused, to ignore whatever lies outside its periphery of attention."[4]

This phenomenon goes a long way toward explaining why Western culture has dominated the field of science for four hundred years. This has worked both as an asset for Western culture and also as a liability. "Because our strength is derived from the fragmented mode of our knowledge and our action,we are relatively helpless when we try to deal intelligently with such unities as a city, an estuary's ecology, or 'the quality of life."[5]

To the extent that the goal of science is knowledge, it must be freed of any direct blame for the environmental crisis. On the other hand, technology, which is oriented to action, derives its power from science and applies it in its projects. According to Max Ways, "Although, its categories are not the same as those of science, technology in its own way is also highly specialised, directed toward narrowly defined aims. As its power rises, technology's 'side effects,' the consequences lying outside its tunneled field of purpose, proliferate with disastrous consequences to the environment—among other unintended victims."[6]

To conclude this train of thought, our argument is that technology is the central agent behind the destruction of the environment, and that the West's innovations in this field derive from attitudes which have been shaped by the dominant religion-philosophical currents of Western culture. The principle of separation or differentiation has produced much scientific discovery and advance, but its very virtue—progress through isolation, individuation, and fragmentation—has also been its weakness in that its tunnel vision has proven myopic in viewing the environment as a whole.

There is one danger in tracing the roots of the environmental crisis to its cultural roots in religion. In his celebrated essay, "The Historic Roots of our Ecologic Crisis," Lynn White asserts that "what people do about ecology depends on what they think about themselves in relation to things around them. Human ecology is deeply conditioned by beliefs about our nature and our destiny—that is, by religion."[7] The danger in this sort of claim is that it makes religion the primary conditioner of human interaction with the environment.

The model which White proposes has three progressive steps: (1) Judeo-Christian tradition, (2) science and technology, and (3) environmental degradation. The picture is more complicated. Certainly religion has a basic role to play, since it accounts for the way we perceive ourselves, our fellows, and the world around us. But Western civilisation has also been conditioned by the institutions of democratisation and capitalism, aided by science and technology. These political and economic forces have produced wealth, individual ownership, urbanisation, and population increase. Each one of these sources has taken more from the environment than it has put in, and thereby has been responsible for degrading it. The principle of separation, the method used by science to separate one aspect of nature from another for the purpose of study, has equally been at work in the operations of our political and economic institutions, and this has only provided us with short-term, piecemeal approaches to our environmental problems. We now turn to the major problems which make up the environmental crisis.

## ECOLOGICAL PROBLEMS

In a rare departure from its tradition of naming a "Man of the Year," *Time* magazine designated Endangered Earth as Planet of the Year for 1988. Like God warning Noah of the coming flood, this was the year that Planet Earth spoke with tremors which were heard around the globe.

In the United States drought stalked the land from California to the Deep South, depleting the grain harvest by 31 per cent, and killing thousands of cattle. The combined effects of heat waves and lack of rain ignited forest fires, with the largest conflagration in Yellowstone National Park. Beaches were fouled with garbage, raw sewage, and medical wastes, showing the limited capacity of the ocean to absorb pollution.

Similar despoliation of the oceans occurred in areas of the Mediterranean, the North Sea, and the English Channel, making beaches unfit for recreation. Killer hurricanes roared through the islands of the Caribbean and sent thousands in Bangladesh to a watery grave. In

Soviet Armenia, 55,000 persons perished in a gigantic earthquake. There were grave warnings of the further depletion of the ozone layer, and all the while tropical forests, which house half the earth's plant and animal species, were being destroyed at the speed of one football field per second. Most of these evils had been going on for a long time, and some of the worst disasters apparently had nothing to do with human behaviour. Yet, this year's [1988] bout of freakish weather and environmental horror stories seemed to act as a powerful catalyst for worldwise public opinion. Everyone suddenly sensed that this gyrating globe, this precious repository of all the life that we know of, was in danger. No single individual, no event, no movement captured imaginations or dominated headlines more than the clump of rock and soil and water and air that is our common home.[8]

With each passing year, threats to the survival of Planet Earth have increased with nothing less than mind-boggling rates. Following are the chief problem areas: biodiversity, global warming, pollution, waste, overpopulation, and the nuclear epidemic.

## THE PROBLEM OF BIODIVERSITY

Mother Nature never puts all of her eggs in one basket: diversity is the key to survival. Variety is not just the spice of life, but its sum and substance. "Complex beyond understanding and valuable beyond measure, biological diversity is the total variety of life on earth."[9]

During the time, when human were hunters and gatherers, their sustenance depended on biodiversity. This dependence changed when society looked for its livelihood, first to agriculture, and then to industry. Now, with the research knowledge that biological diversity is crucial for the environment which sustains life, people are being roused to ensure its conservation.

Our knowledge of the total count of life with which we share this planet is still shrouded in mystery, but what we do know is that biodiversity is vanishing at an incalculable speed. Eugene Linden tells a story of what is happening in Brazil, but which is actually a global phenomenon. Before Brazil's great land rush, the emerald rain forests of Rondonia state were an unspoiled showcase for the diversity of life. In this lush territory south of the Amazon, there was hardly a break in the canopy of 200-ft. tall trees, and virtually every acre was alive with the cacophony of all kinds of insects, birds and monkeys. Then, beginning in the 1970s, came the swarms of settlers, slashir_ and burning huge swaths through the forest to create roads, towns and fields. They came to enjoy a promised land, but they have merely produced a network of devastation. The soil that supported a rich rain forest, is not well suited to corn and other crops, and most of the newcomers can eke out only an impoverished, disease-ridden existence. In the process, they are destroying an ecosystem and the millions of species of plants and animals that live in it. An estimated 20 per cent of Rondonia's forest is gone, and at present rates of destruction it will be totally wiped out within 25 years.[10]

Like Brazil, India is also a land of rich diversity. Its geographic location makes it "one of the world's top twelve 'mega diversity nations.'"[11] But the subcontinent's biodiversity is being dangerously eroded.

One cause for India's loss is earlier British policy, which 'drastically reoriented the resource-use pattern, largely exporting teak, tea, and indigo outside the locality. The prices were greatly depressed. After Independence this system of undervaluing biological resources was taken to extremes. Thus bamboo, of great utility to the rural people, has been made over to the paper industry at royalties of Rs. 1.50 a tonne, when the price for basket weavers was over Rs. 3000 a tonne."[12]

A second cause for the erosion of biodiversity in India is the scant attention paid to the quality of life among its rural and tribal people. "In pre-British India, vast tracts were available to fulfil the biomass needs of the local people. Parts of such lands were set aside as 'sacred groves' for total protection of biological communities, while others were treated as community woodlots from which regulated harvests were carried out."[13] A few of these systems survive today in remote areas such as Mizoram, but they were deliberately and effectively destroyed by the British, who granted no rights but only some privileges to the local communities over common lands. "As a consequence such lands became 'no man's lands' abused by all but protected by none. This situation has worsened after Independence. All this has affected the local people such as the tribals."[14] The stories of India's and Brazil's battles for biodiversity are sadly repeated all over the globe.

Biodiversity is analysed in three ways: by ecosystems, species, and genetic diversity.[15] Ecosystems are the communities in which organisms live and move and have their being. They comprise forests, wetlands, mangroves, and reefs. Of these, tropical forests are of highest value. They cover only 7 per cent of the earth's surface, but they are the habitat for 50 to 80 per cent of its species. Tropical forests are now only half their original size. In 1990, deforestation reduced that aggregated further by an area the size of Washington State.

Of all nations, Brazil has the most extensive range of tropical forests, housing the highest number of species. By 1987, as a result of government action, economic shifts, and climatic changes, the massive deforestation of earlier years was brought under control. "Moreover, with nearly 90 per cent of its groves still standing, by national or international standards the Brazilian Amazon is relatively untouched. Brazil's most endangered ecosystems are its unique coastal forests. Logging and agriculture and urban expansion have destroyed more than 95 per cent of the once-vast coastal rain forests and the coniferous Araucaria forests of southern Brazil."[16] Given the fact that only 5 per cent of the world's tropical forests are protected by their governments, "the stage is set for mass extinctions."[17]

Temperate rain forests are in even greater danger than the tropical forests. Half of the original 31 million hectares have fallen prey to logging. In the United States, only pockets of the old-growth rain forests survive in the Pacific Northwest.

In India, trees are among the most important of natural resources, but this has not lessened the pace of deforestation across the land. In Punjab, forest cover declined from 120,000 hectares to 49,000 hectares between 1972 and 1982,[18] In Rajasthan, some three hundred mines extracting marble, limstone, soapstone, silica, bauxite, and granite have suddenly brought an army of eight thousand labourers into the forests of Alwar district, with predictable damage.[19] Madhya Pradesh is undergoing rapid denudation of its forest cover "from the notional 35 per cent of the area to 19 per cent."[20] In Assam, deforestation is "the greatest threat to the environment in the region as a whole," producing lesser rainfall and severe floods caused by hillside erosion.[21] In Tripura, the Gumti dam project has ousted some, 2,600 tribals from their homes, forcing them to seek livelihood in the upper catchment areas of the Gumti and to resort to the practice of *jhum* (slash-and-burn) agriculture, causing tremendous damage to the trees.[22]

Uttar Pradesh is high among the regions hardest hit by deforestation. "From time immemorial, the Himalayas have played a crucial role in maintaining the ecological balance, particularly in the North."[23] In literature and legend, these pristine heights have served for millennia as the playground of the Hindu gods, and their exquisite beauty has inspired generations of holy men to seek heaven on earth, but like the cedar groves of Lebanon, soon the only record of the Himalyan forests may be found in sacred texts.

Wetlands, like forests, are productive habitats for biodiversity. They are home to countless varieties of fish and waterfowl, and have natural mechanisms which control water flows and cleanse the system of sediments and pollution.

Among India's most picturesque lakes, Renuka, situated in Himachal Pradesh, is a crown jewel. But Lake Renuka is dying. In 1989, the Ministry of Environment and Forests designated it the state's only "natural wetland" and accorded it national status. But the estate's rhetoric has fallen short of conservation efforts to save the lake from being choked by elephant grass and heavy siltation during monsoons.[24]

The greatest loss of wetlands, exceeding 90 per cent, has occurred in New Zealand, Australia, and in California. The blights of heavy industry, urbanisation, agricultural use, over-grazing, and logging are among its enemies. Canada has a quarter of the world's wetlands, which fare the best.

Among coastal wetlands, mangroves are most valuable. The complex ecosystems are good examples of how each part is dependent on the whole for survival. Until recently, no value has been placed on mangroves, and they have been considered destroyable wastelands. As such they have been subject to massive destruction in Asia and the Pacific, where forests are reduced to wood chips utilised by rayon and other industries.[25] India, Pakistan, and Thailand have lost three-fourths of their mangroves. Heavy losses are also recorded for Indonesia, New guinea, Latin America, and West Africa.

Coral reefs are at greater risk than forests or wetlands According to the latest global survey, reefs of off 90 of 109 countries suffer heavy damage due to sedimentation, water pollution, and the effects of fishing and water sports.[26] A ten-year study by the government of the Philippines, where the most diverse reefs in the world are to be found, estimated that 71 per cent of the nation's reefs were in "poor to fair" condition.[27] Along with mangroves, coral reefs are part of an interdependent ecosystem which protect coastal areas from the erosive power of the ocean. They can take the constant beating of the waves, but have little protection against pollution from sewage, fertilizers, and the spread of algae.

Next to ecosystems, species are a second aspect of biodiversity. Scientists say some 10 to 80 million species share the planet with humans. Of this population, 1.4 million species have been identified. But the numbers are fast disappearing. Harvard biologist Edward O. Wilson predicts that 50,000 invertebrate species occypying the tropical rain forests are doomed to die each year, their fate sealed by deforestation.[28] Extinctions take place in nature all the time as a necessary part of the evolutionary process, but man-made extinctions proceed at 1,000 times the rate that has held since prehistory. Peter Raven, director of the Missouri Botanical Gardens, projects that over the next thirty years humans will be responsible for the extinction of one hundred species per day. Forests, wetlands, and reefs are the nurseries for new forms of life, so when these are destroyed, the crisis that looms is the "death of Birth."[29] This represents the gravest assault upon life on this planet, ever since its dawn 4 billion years ago.

It would be tantamount to some maniac going from one hospital to another, blowing up the natal wards.

The invertebrates of the tropical forests are most numerous of species, and by that same calculation have the greatest toll of extinction. Freshwater fish have declined in major areas. In the former Soviety Union, over 90 per cent of commercial fish species have been lost. Similar declines, over a twenty-year period, have been reported for the Atlantic cod and herring, South African pilchard, pacific Ocean pirch, king crab, and Peruvian anchovies,[30] Amphibians have also declined. In India, the state of Maharashtra has suffered the decline of insect-eating frogs, which has resulted in pest damage to crops. The same circumstances in West Bengal have incurred the spread of malaria by mosquitoes, which otherwise would have been controlled by frogs.

The most fecund diversity of plants is also found in the tropics. Because only a fraction of these forests are safeguarded, researchers estimate that a minimum of 12 per cent of the plants in Central and South America are headed for extinction, and therefore must be counted among the "living dead." Four-fifths of the plants that grow in southern Africa are to be found nowhere else; yet 13 per cent of these (2,300 species) are threatened. In the United States, 3,000 species are in danger, with the likelihood that 700 of them mostly natives plants, will vanish over the next decade.[31]

The loss of every species is to be bemoaned because a species is unique and irreplaceable, the repository of millions of years of evolution, a source of scientific knowledge, a group of beautiful objects, and of self-interest to humans, whose well-being is inter-dependent with the lowliest of creatures. Even on a mundane level, humans reap "dividends from diversity" in several practical ways. Biodiversity has especially been beneficial and life-saving to humans in the area of medicine.

In every culture and every age, humans have looked to nature's medicine bag for cures to heal their ills. Queen Nefertiti of Egypt is known to have administered a pain killer in the form of mandrake to her husband. The ancinet Egyptians also used union to cure scurvy among the slaves who built the pyramids. The inhabitants of the Nile use alone to ease intestinal troubles, and henbane as a sedative. Chinese physicians catalogued thousands of herbal cures in the pharmacopoeia *Pen-ts'ao Kang-mu.* Long recognised as a laxative, rhubarb was found growing in China by Marco Polo. Indian doctors calmed disturbed patients with the "medicine of sad men"—rauwolfia, the forerunner of modern tranquilizers. Ancient Greeks and Romans had a vast plant knowledge that included colchicum to treat gout, squill to stimulate the heart, tansy for worms, and fennel and senna for degestion; assassins poisoned their enemies with belladonna, which later provided the medicine atropine.[32] And in modern America pharmacists have used the venom of the Brazilian pit viper to develop Capoten for high blood pressure, among hundreds of other such remedies. All of this is to say that there are very good reasons, involving human survival, to preserve and protect the survival of diversity in plant and animal species.

In addition to ecosystems and species, biodiversity must be viewed at the level of genetic diversity within the species. According to some scientific analysed, the lost of diversity within species poses the most formidable danger to human survival, particularly where food crops are concerned. Farmers have used and created genetic diversity for millenia to increase agricultural production; genetic engineering and other high-tech forms of crop breeding are equally dependent on it. Especially in developing nations, generations of farmers have developed a remarkable array of crops. The Ifugao people of the island of Luzon in the Philippines identify more than 200 varieties of sweet potato by name while farmers in India have planted perhaps 30,000 different strains of rice over the past 50 years.[33]

Ryan senses imminent danger in the tendency in agriculture to concentrate research and production on high-yielding crop varieties because they boost food production at the expense of the many traditional strains that are adapted to local ecosystems and capable of being developed into higher-yielding, locally suitable crops. "At current planting rates, three quarters of India's rice fields may be sown in only 10 varieties by 2005. In Indonesia, 1,500 local varieties of rice have disappeared in the past 15 years, and nearly three fourths of the rice planted today descends from a single maternal plant."[35] The same situation prevails in the United States, where 71 per cent of corn comes from a mere six varieties, and half the production of wheat from nine varieties. The lesson of the Irish potato famine of 1846, which sent waves of immigrants fleeing to the United States is that high levels of agricultural uniformity can suddenly collapse by onslaughts of pests and disease.

Genetic erosion affects the survival of wildlife as well as agriculture. In the United States and elsewhere, deliberate homogenisation has led to the loss of internal diversity among many species, and that poses a threat to their collective survival.

In all of these instances of genetic engineering, humans are engaged in acts which formerly were the prerogatives of the Creator himself. Problems arise, ultimately threatening our own survival (for nothing survives separately), when we ignore nature's own agenda to conserve biological diversity.

## THE PROBLEM OF GLOBAL WARMING

Atmospheric changes that used to take a century are now taking place in a decade. The two main foci of climatic change are ozone depletion and global warming. Though different, both are closely related.

Ozone is a form of oxygen having a molecular structure of three atoms instead of the normal two. Its peculiar structure enables ozone to absorb biologically hazardous ultraviolet 'B' radiation from the sun. The ozone layer is located some 10 to 30 miles in the earth's stratosphere. The vital gas thus serves as a life-support system for planet earth.

By 1972, early warnings were sounded that the nitric oxide produced by supersonic aircraft was damaging the ozone layer. Other causes were subsequently cited, the most formidable being the chlorofluorocarbons (CFCs) from aerosol sprays, refrigerants, and air-conditioning.

On their own, CFCs are inert, but as they rise, solar radiation dissociates them releasing chlorine. The chlorine atoms invade the ozone, removing one of the three oxygen atoms and forming chlorine monoxide. The chlorine monoxide in turn synthesises with another oxygen atom, forming a new oxygen molecule and a chlorine atom. Unlike the ozone molecule, these newly formed oxgen molecules are unable to screen the ultraviolet radiation which passes through to the surface of the earth.

Upon reaching the earth, the effects of ultraviolet light can be devastating. People who are exposed can develop cataracts in their eyes, wrinkling and cancer of the skin, and a breakdown of the immune system. Radiation can affect the growth of crops by interfering with photosynthesis, diminishing yields, and it can impact marine life by affecting the growth of phytoplankton, the chief link in the ocean food chain.

Since the early 1970s, when they were commercialised, vast volumes of CFCs have been released into the atmosphere, the present annual level being some 7,00,000 tons. By 1992, it was estimated that 20 million tons of CFCs had been pumped into the atmosphere.[35] All countries produce these potent chemicals. India's production is at about 5,000 tons per year.[36] What is truly frightening is that once they have been emitted into the atmosphere, CFCs can last for one hundred years or more. This means that even if there were an immediate global ban on CFCs, their effects would still be felt by our great-grandchildren.

But the danger is not in some future; it is here and now. The damages cannot be escaped; the best that can be done is to stabilise ozone loss by the end of this century. Following the confirmation of the ozone hole over Antarctica in 1985, it was thought that Ground Zero was the South Pole. Now, there is evidence that Ground Zero is not just the South Pole. A hole in the earth's protective shield could soon open above Russia, Scandinavia, Germany, Britain, Canada, and northern New England.[37]

Till only recently, there were many doubters of any serious ozone depletion, especially in the Reagan and Bush administrations. Their policies counseled caution until studies could come up with hard proofs. That changed in February of 1992 when the National Aeronautics and Space Administration announced startling findings from atmospheric studies done by a modified spy plane and an orbiting satellite. As the two craft crossed the northern skies last month (Jan. 1992), they discovered record-high concentrations of chlorine monoxide (ClO), a chemical by-product of the chlorofluorocarbons (CFCs) known to be the chief agents of ozone destruction. . . . Previous studies had already shown that ozone levels have declined 4 per cent to 8 per cent over the northern hemisphere in the past decade. But the latest data simply that the ozone layer over some regions, including the northernmost parts of the U.S., Canada, Europe and Russia, could be temporarily depleted in the late winter and early spring by as much as 40 per cent. That would be almost as bad as the 50 per cent ozone loss recorded over Antarctica. If a huge northern ozone hole does not in fact open up in 1992, it could easily do so a year or two later. Says Michael Kurylo, NASA's manager of upper-atmosphere research: "Everybody should be alarmed about this. It's far worse than we thought."[38]

NASA's alarm made the earlier schedule set by the international community to clean up the ozone appear too leisurely. When it was confirmed in 1985 that there was a hole over Antarctica, several nations decided that the problem had to be jointly tackled. In1987, they drew up the landmark Montreal Protocol, which set the goal of 50 per cent reduction of CFCs by 1999. As new evidence of ozone loss augmented, another meeting was held in London in 1990, agreeing to a total ban of CFCS by the year 2000.

NASA's discovery has forced nations to pick up the pace. German Environment Minister Klaus Topfer has issued as call to other nations to match Germany's pledge to terminate CFC manufacture by 1995. In February 1992, the U.S. Senate, by a 96 to 0 vote, considered the situation so critical that they decided to accelerate the phaseout period set by earlier calculations of ozone depletion. The administrator of the Environmental Protection Agency announced that the phaseout could come about by 1996.

The lesson of ozone should be remembered by world leaders as they consider other environmental problems. Business interests are bound to come in the way of environmental priorities because, as the manufacturers of CFCs admit, they are going to entails considerable financially loss. Governments, with an eye to economics, are therefore slow to move in the interests of the people, especial when some of the people are beyond their national boundaries. The standard ploy in such circumstances is to study the subject to death for the declared purpose of learning the truth. Then Senator Al Gore had a point when he said, following NASA's announcement, "Now that there's the prospect of a hole over Kennebunkport, perhaps Bush will comply with the law." If governments always wait until there is metaphysical certitude that a hazard is on hand, it may be much too late to halt the hazard. A good example of bureaucratic foot dragging is the issue of the greenhouse effect, to which we next turn.

The greenhouse effect is a by-product of ozone depletion, but it is potentially more hazardous because it is for more difficult to control. It became a household word when, following the tepid congressional hearings conducted by Colorado Senator Timothy Wirth in the fall of 1987, the topic suddenly generated heat during the summer's disasters of 1988, sparked by the testimony of James Hansen, head of NASA's Goddard Institute for Space Studies: "It is time to stop waffling so much," he declared. "The evidence is pretty strong that the greenhouse effect is here." In point of fact, the summer's drought and record-high temperatures may have had nothing to do with the greenhouse effect and were caused, rather, by random, annual climatic variations, but the events did serve to demonstrate to the government and the American public the climatic calamities that can occur when greenhouse gases are released into the atmosphere.

The greenhouse effect which causes global warming began long before civilisation. The nature occurrence of $CO_2$ gases created a heat-trapping blanket which made life on earth possible. Without this atmospheric shield, the average surface temperature of the planet would be a frigid 0° F instead of 59° F. The mechanics of the greenhouse effect are theoretically simple. The sun radiates energy in the form of lightwaves that are able to reach the surface of the earth because to $CO_2$ molecules in the atmosphere are transparent to visible light, in the same manner that glass panes of a greenhouse allow light to pass through. The heat of the lightwaves is absorbed by the land, water, and organisms. Excess heat is then released back into the atmosphere, not as visible light, but as infrared radiation. Infrared rays are longer and have less energy to escape the atmosphere easily. Some of the rays escape, but the rest are trapped by carbon dioxide and other gases in the troposphere,warming the earth. The amount of heat is correlated to the amount of $CO_2$ in the air.

The problem today is that the industrialised world is emitting vast quantities of its waste products into the atmosphere, including carbon dioxide, methane, nitrous oxide, and chloroflourocarbons. These greenhouse gases prevent additional infrared radiation from escaping, and thus serve to thicken the atmospheric blanket, trapping more heat and warming up the earth even more.

Atmospheric scientists have long known that the earth follows a rhythm of warming and cooling. The present warming phase probably began some eighteen thousand years ago, following the peak of the last ice age. Scientists now believe that these climatic rhythms can be triggered by human activity. Says Stephen Schneider of the National Center for Atmospheric Research: "Humans are altering the earth's surface and changing the atmosphere at such a rate that we have become a competitor with natural forces that maintain our climate. What is new is the irreversibility of the changes that are now taking place."[39]

The consequences of rapid alterations in the atmosphere could be dire. Dramatic weather changes could shift the configurations of deserts and fertile regions, spark tropical storms, cause sea levels to rise as the result of the expansion of water caused by warming, and usher mass starvation and soaring food prices.

An example of these consequences is the plight of the fabled city of Venice. Built on an archipelago in a lagoon, it has long been perched on the edge of disaster. Experts believe that the city has now ceased sinking, but it faces an equally threatening development: the slow rise of the Adriatic, largely as a result of a global warming trend that is causing the world's oceans and seas to expand gradually. Other cities that are thought to be living on borrowed time, affecting some 500 million people, are Miami, New Orleans, Alexandria, Bangkok, Dhaka, Hamburg, Leningrad, London, Shanghai, Sydney, and Thessalonika. Closer home, Waikiki could well be the Venice of Hawaii. Other isalands of the Pacific could face similar fates. The Marshall Islands are in particular danger, because many of them top off at 12 feet above sea level. Global warming could be sending nations, ancient cultures, and ecosystems to watery graves.

But the current data are insufficient to make specific predictions. Notwithstanding the high-level production of carbon dioxide over the past two hundred years of industrialisation, it has not yet been fully proven that humans have triggered a process of global warming. It is easy to mistake short-term variations like droughts for long term climatic change. The picture will only become clearer toward the end of the 1990s. The Earth has great capacities to adjust to changes. A warmer ocean could create more cloud cover that could cool the earth and thus offset the greenhouse effect. It is also possible that the ocean could absorb the carbon dioxide in the air.

There is much we do not know, but on the basis of what we do know, it would be imprudent to flirt with disaster on a global scale. There yet may be no conclusive evidence linking human actions to the greenhouse effect, but this does not mean that sometime in the future the problem is not going to materialise. Researchers may differ over the temperature effects of the carbon dioxide build-up, but no one disagrees about the carbon dioxide build-up itself. Besides, time is not on our side.

Even if the build-up of carbon dioxide and other gases does not produce significant temperature rise, researchers such as Harvard University's Fakhri Bazzaz are warning that the increase could disrupt farms and forests. The main culprit behind the greenhouse effect is the use of fossil fuels by industrial nations. It is too much to expect that any significant changes in the status quo will be ushered in before the middle of the next century, because there are high economics takes, backed by the moghuls of the oil, coal, and auto industries.

## THE PROBLEM OF POLLUTION

In the spring of 1989, the *Exxon Valdez* ran aground on Bligh Reef, Alaska, and spilled 260,000 barrels of crude oil into one of the world's most pristine bodies of water. It cost Exxon a billion dolars for the cleanup, and the state of Alaska millions more to restore the beaches fouled with oil. But the cost to wildlife exceeds a dollar value: more than 34,000 birds and 984 sea otters fell prey to the slick, among other animals.

In the spring of 1991, Iraq's "scorched sand" tactic in Kuwait brought a brutal twist to the Gulf War. Saddam Hussein's retreating army set ablaze five hundred oil wells (roughly 50 per cent of the nation's total). Thick streams of soot darkened the sky over more than a third of the Kuwait battlefield and beyond into Saudi Arabia. The desert was suddenly turned into a burning wasteland. One soldier said it all: "It's amazing. You look at Kuwait, at the whole area. It's just fire." In the light of the burning wells, and under the pall of black rain, there were fears of ecocalamity, but in the months that ensured, the scars of Saddam's scorched-earth policy against Kuwait proved terrible but not catastrophic.

These recent environmental disasters join the ranks of Love Canal, Chernobyl, and Bhopal to dramatize man-made environmental pollution that threatens the health, not of humans alone, but of birds, beasts, and beaches. Sometimes the threat is visible, but mostly it is silent, pervasive, deadly, and a growing menace to the health and safety of people everywhere. The soot generated over Kuwait had soon travelled across to Asia, so though we examine particular instances, we are also aware of their global impact.

Until recently, it was thought that disease occurred spontaneously, but there is growing evidence that many diseases are environmentally induced. We shall try to identify that connection, confining our remarks to the connection between pollution and two medical problems: chronic respiratory disease and cancer.

Our data are derived from the Environmental Protection Agency's 1987 inventory of toxic chemicals emitted by U.S. industries.[40] The EPA's inventory is a result of a 1986 federal reporting law that covers a list of 328 hazardous chemicals and compounds. The Emergency Planning and Community Right-to-Know Act mandates that large industries report their annual chemical use. The regulation followed on the heels of the chemical disaster in Bhopal, India, that killed and injured thousands of people.

The EPA's inventory reveals a gargantuan problem generated by industrial emissions of toxic pollutions. In 1987, industries spewed out 7 billion pounds of toxic chemicals, with 2.65 billion pounds being released into the air. Industries are making the skies into a garbage dump. Among the fifty state, Texas ranked first in the amount of toxic chemicals. Comfort,Texas, released 465 million pounds of toxic wastes; most of this was buried, but the numbers are not very comforting to the neighbourhood. Hawaii ranked last, pumping 2.6 million pounds of toxic chemicals into the air, land, and water.

The EPA toxic inventory is a step in the right direction. For the first time, the public has some idea of the national pollution problem and known that it has a "right to know". However, the inventory is marred by numerous limitations and loopholes. Only large plants processing 75,000 pounds of toxic chemicals per annum are required to report. The statistics do not cover car emissions, power plants, the military, municipal sewage, or small companies. We, therefore, are left in doubt about the toxic total. There is no doubt that whatever the true numbers are, environmental pollution is responsible for chronic respiratory disease, cancer, birth defects, mutations, and psycho-behavioural defects.[41] We take a closer view of chronic respiratory diseases and cancer.

Large numbers of Americans suffer from bronchitis, asthma, an emphysema, which are caused by air pollution, especially from the sulphates found in auto emissions. Samuel S. Epstein states: "It has been estimates that failure to meet current sulphur dioxide standards, which are exceeded in most U.S. cities, is responsible for six thousand premature deaths annually, six to ten million avoidable asthma attacks, and twenty to thirty million days of exacerbated cardiovascular and respiratory disease."[42]

A host of other chemicals are hazardous to the lungs and contribute to chronic respiratory diseases, including ammonia, a colourless gas used as a refrigerant and in the manufacture of fertilizers, synthetic fibres, dyes, and explosives; chlorine, a heavy greenish-yellow gas used for water treatment, food processing, disinfectant production, and cleaners; and sulphuric acid, a colourless, odourless liquid or crystal used in soaps, auto batteries, animal feed, and polishes. Occupational dust is also a toxic agent which causes various forms of pneumoconioses. An estimated 125,000 coal miners are afflicted with this disease, which enfeebles the lungs and kills approximately 4,000 workers each year.[43] This debilitating disease takes other forms as well: asbestosis from asbestos dust; byssinosis from cotton dust; bagassosis from sugar cane dust; and silicosis from silica dust. Every year in the United States some 9,000 deaths can be traced to these dust-related occupational diseases.[44]

The Greeks called cancer *karkinos,* "the crab-like disease," because of its spreading growth. In its Latinised form, *cancer* is the most ominous word in our modern vocabulary because it is seen as a death sentence. Although, cancer is second to cardiovascular disease as a cause of death, it is the most feared. Cancer afflicts 1 in 4 Americans now living, with approximately 800,000 new cases diagnosed each year.[45] One-third of those will survive to live out a normal span of life, but such statistics do not diminish the fears.

Cancer is not a single disease but a combination of different diseases identified by their unregulated cell growth. Therefore, the causes, diagnoses, and treatments of cancer vary. In the past decade, the media have spotlighted the carcinogenic properties of household items such as saccharin, cyclamate, and methpyrilene (used in sleeping tablets), known from laboratory experiments on rodents. We do not yet know whether these chemicals cause cancer in humans, but research sheds new light on the rôle of chemicals in producing cancer. Until only a few years ago, it was thought that cancers were caused by hereditary factors, and perhaps viruses. Researchers are now examining the role of natural and synthetic chemicals in the production of the majority of human cancers.

The evidence for environmental sources comes from various directions. First, although different cancers are variously distributed in different parts of the world, cancers occurring in the families of immigrants to another country tend to take on the cancer pattern of the adopted homeland. Second, it is now documented that exposure to certain chemicals in the workplace causes cancer. For instance, the known carcinogen benzene is associated with leukemia. It is used in the manufacture of rubber, detergents, and dyes. Asbestos has been a much-discussed work hazard in recent years. A mineral founding rocks, it is a durable product found in construction and industrial work. The danger comes when the absestos fibres break into microscopic dust that is inhaled by the workers, producing a condition known as *mesothelioma*, or cancer of the chest and abominal membranes. The disease is incurable. Third, the most convincing data linking cancer to chemicals in the environment arise from the high mortality rate of lung cancer. During the first fifty years of this century, American males increased their smoking by 10 per cent and their death rate from lung cancer rose 100 per cent; the cycle has been repeated by American women since the end of world War II. In the words of one cancer researcher: "It is almost as if western societies had set out to conduct a vast and fairly well-controlled experiment in carcinogenesis bringing about several million deaths and using their own people as the experimental animals."[46]

There is growing consensus that, although viruses do produce cancers in laboratory situations, "they seem unlikely to be primarily responsible for causing any of the common human cancers." Although heredity does play a part in increasing the susceptibility to cancer, there is not hard evidence that common tumors are inherited.[47] The spotlight therefore shifts to environmental Chemicals as the causes of cancer. These chemicals may be synthetic, produced in factories, natural chemicals found in homes (food contaminants), or chemicals produced by bacteria in the large intestines.

The pollution problem in the United States in replicated in the rest of the industrial world. History may yet prove that than real damage done to Eastern Europe was not political or economic, but environmental. Communist leaders were wont to proclaim that pollution was a by-product of capitalism, but it is now an ironic discovery that political systems that encouraged profits, like the United States, proved more resistant to environmental exploitation than systems that disavowed profits. Under the socialism of Karl Marx, natural resources were to be factored into the cost of production, so the present environmental holocaust should theoretically not have occurred. But there was a big gap between theory and practice. In practice, the individual was reduced to a cog in the political machinery. and talk abut the "right to know" would have been deemed subversive. People were prevented from protesting known catastrophes, and steps were taken to keep information on pollution, secret. The build-up of heavy manufacturing, begun by Josef Stalin after World War II, had no impetus to improve its technology in the absence of competition. The more outdated the technology, the greater its capacity to pollute. Similarly, the demands for more food were met by greater distribution of free fertilisers to farmers. The consequence: "contamination by cancer-causing nitrates that will take decades to cleanse from the water and soil."[48]

Now, with the lifting of the Iron Curtain, the world can see the extent of the deadly pollution in Eastern Europe. We concentrate on the area where Poland,

Czechoslovakia, and East Germany meet, for these are "the planet's most polluted lands."

Czechoslovakia is the worst polluted. Seventy per cent of rivers "cannot be used for drinking, fishing or swimming."[49] Only six rivers in Bulgaria are fit for human use. The Danube is blue only in song: "it flows brown and polluted at every point on its 2,000-mile journey to the Black Sea." Among the new "dead seas" are the Caspian, Baltic, Barents, Black, and White seas. Air pollution has defaced statuary and stone buildings, so there is no doubt what it has done to the health of people. Pollution, stemming from intense coal mining and burning, has produced acid rain that has replaced once lush forests with miles of dead trees.[50] Josef Vavrousek, head of Czechoslovakia's new environmental agency, could well be speaking for all of Eastern Europe: "We are an example to the world of how not to treat the environment."[51]

The same can be said of the former Soviet Union. In January of 1991, then President Mikhail Gorbachev said in a speech: "The time is ripe to set up an international mechanism for technological cooperation on environmental protection."[52] The need to reverse decades of pollution could hardly have been more urgent. Some 50 million Soviet citizens were living in environments where the pollution levels were ten times those prohibited by the West. The estimated costs for tackling the pollution problems over the next ten years was a hefty $40 billion annually,[53] the Chernobyl disaster defies all estimates.

In South America, the worst-case scenario of air pollution is Mexico City. Morning, noon, and night, 23 million souls breathe in the vortex of a toxic cloud. The winters of 1991 and 1992 were particularly severe. The people call it *nata,* or scum. It hangs like a brown, rotting mushroom over the megapolis, composed of carbon monoxide, sulphur dioxide, nitrogen dioxide, and ozone. It has produced a 16 to 20 per cent rise in respiratory infections, nose-bleeds, and emphysema. The high volume of noxious gases that is daily pumped into the air makes it impossible for the atmosphere to recuperate, creating a situation that is chronic. A full 7 per cent of the city's industrial pollution was being caused by an oil refinery which spewed 88,000 tons of contaminants into the atmosphere every year. It had to be shut done, at great financial loss.[54]

The developing countries of Asia present new sources of pollution which could surpass those of the industrialised nations. Two opposite forces are simultaneously at work. On the one hand, there is the vertical thrust as the developing countries try to catch up with the more affluent West. To find their place in the sun, they are pushing their factories and farms to the limit. On the other hand, there is the horizontal counterforce of explosive populations. With more mouths to feed, they must push even harder, sacrificing health for economic gain. Pollution control is a luxury they can ill afford. In the bargain, the sun sets covered with smog, disease spreads, and the price of success often exceeds economic gain. It is a vicious circle, and even when it is medical knowledge that the health stakes are high, poverty makes industrial slaves.

An example is the Kesoram Rayon Factory at Tribeni, near Calcutta. A report states that, without warning, the factory emits an invisible poisonous gas which soon seeps into the homes of the fourteen villages in the immediate neighbourhood. "Coughing, retching, the suffocated victims run futilely out of their homes in search of fresh air."[55] For twenty years the factory has been contaminating the air with sulphur dioxide, sulphur trioxide, and carbon disulphide gases. A spokesman for the villagers says, "We have been fighting against the

Kesoram factory management since 1961 to prevent their gases from destroying our lives."[56]

In response to picketing of the factory, the management has made promises to improve things, but "typically, nothing was done." The environment minister was invited to study the situation. "The visit achieved nothing, nor has the state Environment Department been able to do anything. Curiously enough, the environment minister's son....has been given a job as a chemist in the factory."[57] The villagers have given up organising committees to lead the agitation "because in the past, committee members have been silenced by being employed in the factory."[58]

There is no denial on the part of the factory management that they are emitting poisonous gases. "We have two sulphuric acid plants with a total capacity of 50 tonnes per day. During restarting, a huge quantity of sulphur dioxide gases escape as stack gas which are not only losses in production, but they also pollute the atmosphere of the neighbourhood."[59]

The health hazards of these gases are well known, Sulphur dioxide attacks the respirator system, causing bronchitis and other complications such as gastritis, and eventually coronary diseases. Carbon disulphide is worse. It damages the nervous system, causing behavioural impairment, including the loss of sex drive. The gases also destroy plant life, killing the bamboo and mango trees that otherwise thrive in the area.

The report ends on a tragically realistic note: "Understandably, despite everything, nobody wants the factory shut down. Says schoolteacher Meera Chakraborty: 'How can we demand the closure of the factory when the livelihood of so many depend on it?' Pollution control equipment could minimise the harmful effects of the gases, but it is expensive."

B.B. Sundaresan declares that the problem of industrial pollution is an all-India problem. "The rapid growth of industries, especially chemicals and allied products, has resulted in the production and use of substances some of which are health hazards. A significant amount of these compounds also get released into the environment, affecting the flora and fauna. A number of organic chemicals are being synthesised and their toxicity values are so profound that they are gradually threatening the environment."[60]

Asian countries that do make it to the top often find that they are victims of their own success. Taiwan is a country that is rich in cash, and also rich in pollution. It is now as cash rich as Japan, with $70 billion in foreign-exchange reserves. In the forty years, since, the Nationalist Chinese fled here from the mainland, Taiwan has burgeoned from a small backwater into a "major economic powerhouse. Heavy industry has been the strength of its economic muscle. But the price has been high. Rigorous development programmes have spawned industries which are among the worst polluters in the business. Says K.L. Chang, professor of physics at the National Taiwan University, "When other countries were phasing out these industries in the 1960s, were welcomed them with open arms. Now we're stuck with the pollution."[61]

According to one environmental report, Taiwan is one of the dirtiest places in Asia. At least one third of the island's rivers and one eighth of its farmlands are contaminated. Air quality in parts of Taipei routinely surpasses the harmful level, and visitors to the city often develop a lingering cough. Exhaust fumes from cars, motorcycles and scooters leave a permanent haze over the capital. In the countryside, households and villages dispose of garbage and refuse by dumping it into a field or ravine. Asthma cases among Taiwanese children have quadrupled in the past 10 years. Cancer, Taiwan's leading cause of death, has doubled over the last 30 years. And although the Government has taken modest steps toward safeguarding the environment, state run enterprises and large corporations continue to pollute with impunity.[62]

As Edgar, Lin, former president of the Taiwan chapter of Greenpeace, sees it: "There is no political will to change. The situation keeps getting worse. You can see it, feel it and breathe it."[63]

Japan reveals yet another side of the face of Asia. Japan leads the world in the manufacture of fuel-efficient automobiles. With China, it is one of the largest consumers of gasoline in Asia, and the fuel is largely lead free. Japan also has stringent specifications for sulphur in oil products in order to combat air pollution and acid rain.

By all accounts the engine which drives the Japanese economy is finely round, but within the mechanical complexity, there are cultural components which must not be overlooked. The cultural elements show that Japan has a somewhat split personality in regard to the environment. Professor Stephen Kellert of Yale University explains that the Japanese "display a detached appreciation for the beauty of species that fall within their cultural and historical heritage and can be manipulated to fit within their stylised picture of what the natural world should be."[64] On the other hand, "nature in the raw" is often treated as an alien force to be overcome. In this role, Japan was recently criticised by *Newsweek* as a "whale-killing, forest-stripping bogeyman on the environmental stage."

Alan Miller, a specialist on Japanese environmental policy and executive director of the University of Maryland Centre for Global Change, views this attitudinal schizophrenia contextually. By one report: He believes the Japanse people's split environmental personality stems from their island home's lack of significant national resources and the absence of a wilderness tradition.

For centuries, the predominantly agrarian Japanese society had to battle the elements to eke out a living. Being inhabitants of an island nation, the Japanese viewed marine life as a commodity to be exploited, not as anything possessing an intrinsic value of its own. The sea was a force to be feared, not revered.

On top of all this, Japan made a decision that it had to industrialise in order to compete with the Western World.

Economic growth became the secular religion, with environmental concerns not even worthy of an after thought.

This was reflected in modern Japan's environmetally insensitive overseas trade policies, particularly its participation in and encouragement of overharvesting of tropical forests for commercial gain.

But even on the domestic front, where many of the Japanese government's pollution control efforts have been relatively effective, the remedial measures have tended to be localised and quite limited.

The environmental movement in Japan remains embryonic. There is little if any public participation in environmental decision-making. No citizen environmental organisations of any consequence are around to badger authorities to act responsibly.

Miller and other expert observers of Japan attribute this set of circumstances to the nation's ancient tradition of placing blind faith in establishment authority.[65]

At the same time, there is hope for positive change because, on the same cultural note, it should be observed that Japan is sensitive about how it is being perceived by the world, especially at this time when it is developing a strategy for world leadership. The West has the chance to influence Japan's environmental policies, and recently, Japan has bowed to changes demanded by Green-peace and other agencies.

The most lethal form of pollution derives from nuclear power. It is the stuff of doomsday books, summoning the fears of a meltdown at Three Mile Island, and the hellish explosion at Chernobyl, rocketing a cloud of radioactivity over the Ukraine and Europe.

Eduard Shevardnadze takes us back to that red-letter day, now become a watershed in environmental thinking. On that morning, the telephones jangled at an hour when they are normally silent. I postponed my usual Monday meeting to answer all the agitated questions. They all boiled down to one: What had happened? I had no answer. During those first morning hours I didn't yet know that an accident had occurred at one of our nuclear power stations. My efforts to extract news from the sources I could reach produced little. "Yes," they told me. "Somethings happened in the Ukraine, but so far we can't get any clear information." Scraps of news trickled in, but did not give the full picture of what had happened. Gorbachev telephoned and asked me to come right over. By that time, more than a dozen foreign ambassadors had requested emergency meetings with me or my deputies. Their governments had urgently ordered them to find out which radioactive elements were spewing into their countries' atmosphere, soil and water. Once they had ascertained that their own reactors were functioning normally, they had traced the leak to the Soviet Union. The affair was turning into a scandal![66]

The then foreign minister goes on to state that though he had been raised in a spirit of mystical veneration for the "top secret" stamp, in this instance, the principle of *glasnost* had to displace all ideological safeguards, and the whole truth had be told to the world. "Chernobyl Day," as I privately called April 26, 1986, marked a watershed in world history, a new criterion for foreign policy. We hadn't managed to pronounce this phrase yet and had not clarified the real dimensions of the catastrophe, but in one second, the explosion heralded an era when no ecological disaster caused by a breakdown in technology could be kept within rational borders. In the modern world, the boundaries and radius of various natural disasters were very conditional. There was an unconditional need for outside international resistance to a threat no less than a thermonuclear war: the ecological collapse of a planet.[67]

Shevardnadze concludes his remarks on a new note. "'Chernobyl Day' tore the blindfold from our eyes and persuaded us that politics and morals could not diverge. We had to gauge our politics constantly by moral criteria."[68]

In spite of the new policy of *glasnost,* Soviet official secrecy prevailed. Now, it is known that the true body count of that fateful day was as many as ten thousand dead by the explosion, and millions more afflicted by its radioactive fallout. Gorbachev marked the fifth anniversary with the warning that "humanity is just beginning to realise fully the global nature of social, medical and psychological problems created by the catastrophe."

Back home, the message of Chernobyl is being used by antinuclear groups, such as Greenpeace, to discontinue use of nuclear power for domestic energy. In 1992, the United States derives 21 per cent of its electricity from nuclear power, as against 27 per cent for Japan, 33 per cent for Germany, 36 per cent for Bulgaria, 60 per cent for Belgium, and 75 per cent for France.

On the credit side, whereas fuels (coal, oil, and gas)

produce greenhouse gases and other pollutants in large quantities, nuclear energy is a power source that is clean. It is derived from plentiful uranium, and the new generation of reactors and waste disposal are increasingly safe. On the negative side, nuclear fuel and waste are highly radioactive. As a general comment: nuclear power is a vast energy source, immensely oversold at the beginning, but is now making a comeback.[69]

## THE PROBLEM OF WASTE

The notion of the world as a global village is fast being replaced by the reality of the world as a global wasteland. Throwaway societies are befouling their land, air, and oceans. Believing that the planet has an infinite capacity to digest the volumes of waste we feed into it, no nations has ever thought this to be a problem. The common wisdom has always been: "Out of sight, out of mind." Now, we are down in the dumps. The urgent global dilemma facing America and the rest of the world is: "how to reduce the gargantuan waste by-products of civilisation without endangering human health or damaging the environment."[70]

A symbol of our civilisation was the two-year odyssey of the freighter *Pelicano* which, like the legendary *Flying Dutchman,* seemed doomed to sail the seven seas endlessly, with no ports willing to receive her cargo. And with good reason. The ship was carrying 14,000 tons of toxic ash that was loaded in Philadelphia in September of 1986. In October 1988, the captain impudently dumped 4,000 pounds of its unwelcome cargo in Haitian waters. A month later he struck a deal with an unnamed country to accept the noxious load.[71]

As John Langone observes, the voyage of the *pelicano* typifies the "environmental exploitation of poor countries be the rich," and also "the single most irresponsible and reckless way to get rid of the growing mountains of refuse, much of it poisonous, that now bloat the world's landfalls. Indiscriminate dumping of any kind—in a new Jersey swamp, or a Haitain beach or in the Indian Ocean—simply shifts potentially hazardous water from one place to another."[72] The common mentality behind this peculiar saga is the NIMBY mentality—the acronym for Not In My Back Yard.

The problem of waste is one that no one wishes to see, let alone smell. Yet the global wasteland covers all nations, rich and poor. From the festering industrial landfills of Bonn to the wastechoked sewage drains of Calcutta, the trashing goes on. A poisonous chemical soup, the product of coal mines and metal smelters, roils Polish waters in the Bay of Gdansk. Hong Kong, with 5.7 million people and 49,000 factories within its 400 sq. mi., dumps 1,000 tons of plastic a day—triple the amount thrown away in London. Stinking garbage and human excrement despoils Thailand's majestic River of Kings. Man's effluent is more than an assault on the senses. When common garbage is burned, it spews dangerous gases into the air. Damped garbage and industrial waste can turn lethal when corrosive acids, long-lived organic materials and discarded metals leach out of landfills into ground water supplies, contaminating drinking water and polluting farmland.[73]

America's affluence and industrial might also gives it a high place on the waste heap. We are familiar with the grim statistics of the nation's trash: We generate more than 160 million tons of solid waste a year—approximately 3.5 pounds a day per person, or 1 ton each year. "A convoy of ten-ton garbage trucks carrying the nations' annual waste would reach half way to the moon."[74]

- * We create one ton of solid industrial waste every week for each citizen.

* We produce an average of 20 tons of $CO_2$ per year for each individual.
* We discard enough aluminum cans power annum to rebuild the whole U.S. commercial airline four times over.
* We cost off sufficient iron and steel every year to supply the total needs of GM, Ford, and Chrysler.
* We toss out each year enough writing paper to erect a 12-foot wall, reaching from Los Angeles to New York City.
* We throw away every two weeks the number of glass bottles and jars it would take to fill both Towers of the World Trade Centre.

Between 1960 to 1986, the volume of American garbage grew 80 per cent. By the turn of the century, the malodorous mound is expected to rise by 22 per cent to a volume of 192.7 million tons.[75] Only the Australians come close to U.S. statistics, while the West Germans and the Japanese come in at half the figures.

Among Americans the most profligate offenders are Californians, who throw away an average of 2,555 pounds a year. Says one environmentalist: "In Los Angeles County, we generate enough trash to fill the Dodger Stadium with garbage every nine days or So."[76]

But Dodger Stadium is not prepared to provide this service, and the number of other less prestigious facilities is fast diminishing. This is what is posing the present waste problem. Says Vice President Al Gore: "the volume of garbage is now so high that we are running out of places to put it. Out of the 20,000 landfills in the United States in 1979, more than 15,000 have since reached their permanent capacity and closed. Although, the problem is more acute in older cities, especially in the Northeast, virtually every metropolitan area is either facing or will soon face the urgent need to find new landfills or dispose of their garbage by some other means."[77]

The epicenter of the crisis is New York City, which produces 44 million pounds of garbage every single day. Mountains have always served to inspire and elevate the human spirit. And, then there is "Mount Trashmore" on Staten island, rising, load by solid load, toward a peak of 505 feet. Beyond that point there would be the jeopardy of a garbage avalanche. It will have to obtain a permit from the Federal Aviation Administration, posing danger to air traffic. Beside this "Everest" of paper, aluminum, tin, glass, motor oil and car batteries, other monuments are planned for heights of 450 feet and several at 300 feet. The mounds will be covered with clay and soil, making, parts of the areas into a park. Probably, some entrepreneur will come along and change all that into a ski resort.

Eighty per cent of the nation's waste is buried in landfills. In the public perception, landfills appear "natural" because we are told that the waste in them is biodegradable. Landfills are therefore viewed as huge compost piles. In fact, they should be seen as solid vaults which preserve the garbage because it is covered, compact, and shielded from sun and moisture. The slow decomposition that does take place is an efficient producer of methane gas that contributes to the greenhouse effect.

In 1991, the Environment Protection Agency set up new standards to ensure that the country's remaining six thousand landfills be guarded against hazardous leakages of gases and pollutants. This will add some 900 million dollars to the nation's waste costs, now calculated at $4 billion to $5 billion annually.[78]

There are a few alternative forms of garbage disposal.

Incinceration reduces the weight by 70 per cent, and produces heat that can be sold to generate electricity. Shipping wastes across state lines is commonly practised, but southern states seem to be mobilising against the invasion of Yankee garbage. Third World countries also resent the "garbage imperialism" of rich industrialised nations. Gore reports: On the West Coast, some municipal officials in California have been negotiating with the Marshall Islands in the South Pacific to receive regular shipments of solid waste. The residents of these islands, many of whom are suffering from the lingering effects of the U.S. government's atmospheric nuclear testing programme in the 1950s, would not ordinarily consider receiving such an unsavory and even dangerous import, but their poverty forces them to do so. Meanwhile, Greenpeace recently disclosed that officials in Baltimore were negotiating with authorities in China for permission to dump tens of thousands of tons of municipal solid waste in Tibet. Nothing could be more cynical. The Tibetan people are powerless to prevent Chinese officials from destroying the ecology of their homeland because of China's armed subjugation of Tibet for the last forty years. But the shipment has not taken place, and the United States has not yet become heavily involved in cross-border waste trafficking.[79]

The most environmentally sound way to tackle the waste problem is through recycling. Japan recycles 50 per cent of its trash; Western Europe, 30 per cent; but the U.S. only recycles 10 per cent (16 million tons per year). Cities such as Seattle, Newark, San Francisco and San Jose are leaders in the field. What impedes the process is that the way goods are packaged for the market place makes recycling expensive and complicated. Besides, there is lethargy on the part of the public. "Many Americans simply refuse to be bothered with sorting and separating garbage into recyclable and nonrecyclable parts."[80]

The waste problem is complex, but its should be plain that the element of culture has much to do with it. As the privileged citizens of a wealthy nation, we Americans have developed a throw-away, convenience culture which is responsible for this mess. Before, we alter our production processes, and radically reduce the quantity of waste we produce, we must dump the cavalier way in which we think about waste.

## THE PROBLEM OF OVERPOPULATION

Overpopulation is the ultimate environmental problem. Its sudden surge in our century defines and dramatises all aspects of our ties to the earth. Each of the problems we have looked at can be reduced to the numbers of the people on this planet, at any given time and place, except that now time and place have vanished, and nations are scrambling for what remains of earth's precious resources.

To illustrate our point, let us examine a single resource: water. Whatever high principles the Gulf War of 1991 was fought over, it finally boiled down to oil. That era is ending. More vital environmental issues are now at stake. Future wars will be fought over water. The thought is not that far-fetched when one considers the claims and counterclaims to rivers and lakes being made by Israel and Jordan, Egypt and Ethiopia, and India and Bangladesh, among others. Fourteen countries depend on the Nile for their water supply, and all fourteen of them are bursting at the seams; yet the Nile has no more water than in the days of Moses. Similarly, the Tigris and the Euphrates and the river Jordan are no better than in biblical times, but their limited supplies are in demand by competing groups of growing populations. Professor Arnon Sofer, a geographer at the University of Haifa, sounds an alarm: "Wars over water might erupt in the Middle East in the '90s when the states try to control each other's supplies."[81]

Water is an issue because population growth and development have depleted and polluted the world's water supply, raising the risk of starvation, epidemics, and wars.

The problem of overpopulation is best told by the figures themselves. It took 3 million years for the world's population to reach the billion mark in 1830. A hundred years later, it hit 2 billion in 1931. Thirty years later it reached 3 billion in 1960. By 1974, in fourteen years, it jumped to 4 billion. In 1986, it was 5 billion. By 1996, it will be 6 billion. Demographers had been predicting that the world population would stabilise at 10 billion in the next century, but now that figure is revised to 14 billion. But statistics canbe abstract. More concretely, every decade the world is adding one China to its population; every year one Mexico; and every month one New York City.

More disturbing is the fact that 90 per cent of population growth is taking place is Third World countries. In Africa, the numbers will rise from 650 million to 900 million by the end of the century. In Latin America and the Caribbean, the figures will increase by 100 million to 540 million at the turn, of the century. A rise in China's birth rate and an apparent halt to the decline of India's fertility rate are major factors in the accelerated growth of population. Together, China (1.087 billion) and India (816.8 million) make up 37 per cent of the world population.

A contributing factor to the world's population growth is improved life expectancy, now 63 years worldwide, up from 45 in 1950.

The figures are multiplying fastest in urban areas. Lee-Jay Cho, Director of the Population Institute, East-West Center, Honolulu warns: The increasing urban concentration is an important aspect of demographic change affecting both rural and urban development efforts. Urbanisation in the next two decades will bring about drastic changes in socioeconomic, and political life. The increasing number of people in cities thwarts governmental efforts to improve the quality of life for current and future generations. When such growth becomes unmanageable—that is, when jobs and public facilities lag behind the demand for them—then urbanisation becomes a serious concern for public policy.[82]

By any measurement, the impact of overpopulation on the environment is frightening. In Africa, we are already seeing the spread of new epidemics such as cholera, the black plague, and AIDS. Millions are starving to death. In Asia and Latin America, we witness the rampant denudation of forests, missives soil erosion, and the depletion of natural resources,which are putting a critical strain on the integrity of their ecological systems.

The intractability of the population problem lies in the fact that even if fertility rates in the Third World declined from four births per woman to the replacement level of two, the population would go on climbing in the next century "because the vast numbers of females born on the steepest part of the S curve in the '50s and '60s have generated 'demographic momentum,' a boom in childbearing that will last for some time to come."[83]

How big that boom is and how long it persists is uncertain It would be tolerable if the world population would stabilise at 10 billion rather than 14 billion. At double the present population, humans will still be able to feed themselves and prevent destructive social, political, and environmental consequences; but at three times the present population, one might fear the possibility of a Malthusian cataclysm.

North American and European countries (with the

exception of Albania) that have growth rates of less than 1 per cent should not consider themselves as somehow separate from the problems of overpopulation. Though Western population rates may be low, our consumption rates are high, and these are only being satiated at the expense of poor countries, which must trade their precious resources for us to maintain the standards of affluence to which we have become accustomed.

Moreover, because we share a global economy, the increase in numbers in a Third World country is felt worldwide. For example, massive unemployment in Asia and Africa affects the international economy, making it harder for the West to compete with Third World labour markets.

Most ominously, according to a UN report, *The state of the World Population 1991,* migratory pressures will increase as the Third World population rises. "The number of migrants fleeing poverty, political oppression or environmental instability already is growing while the need for unskilled and semiskilled workers is falling in the industrial nations."[84] The phenomenon of the boat people risking all to reach Hong Kong, only to be returned to Vietnam; illegal immigrants from Pakistan, Sri Lanka, and Nigeria being attacked by "skin heads" and neo-Nazis in the streets of Germany; and the rickety armadas of Haitian refugees being shipped back from American ports to their homeland augur more desperate times which developed nations will find more difficult to overlook.

The possibility of such catastrophes occurring was predicted by Aldo Leopold in his classic conservation primer, *A Sand County Alamanac.* According to his studies of deer, when the natural ememies of these animals is removed, their populations grow rapidly. Then, because of crowding and competition for limited space and food, their numbers are decimated by starvation and disease.

Leopold reveals a grim similarity between the graphs of the rise in the deer populations and in the present growth of human population. Outcomes are no different for man and beast. The crisis is already bringing the beast out of men.[85]

Religion is at the core of the population crisis. "Of all entrenched values, religion presents perhaps the greatest obstacle to population control Roman Catholics have fought against national family-planning efforts in Mexico, Kenya and the Philippines,while Muslim fundamentalists have done the same in Iran, Egypt and Pakistan."[86]

U.S. contributions to population control diminished after President Ronald Reagan cut support for family planning programmes from $290 million in 1985 to $230 million in 1988. The slash signified his administration's opposition to funding programmes involving abortion. When George Bush was the U.S. representative to the United Nations, he wrote in 1973 that "success in the population field" might "determine whether we can resolve successfully the other great questions of peace, prosperity and individual rights that face the world."[87] As president, Bush switched to Reagan's policy out of fear of Republican rightwingers who proclaim they "pro-life." In the realities of the Third World, that slogan veils a policy that is pro-death. Now with President Bill Clinton in office, the pre-Reagan U.S. policies on family planning are being restored.

## THE PROBLEM OF THE NUCLEAR EPIDEMIC

For forty years, the world lived with the constant threat of a nuclear holocaust. Suddenly, with the collapse of the Soviet Union, the world breathed a sigh of relief that the prospects of a doomsday had now ended. The United States was now the only military superpower, and it could be counted upon to keep the world safe from incinerating itself. That reprieve was short-lived. By early

1992, a more ominous threat to planetary survival mushroomed in the form of "the nuclear epidemic." Cover stories in our national magazines shocked us with the news that "the West's attempt to prevent the spread of nuclear weapons has failed, and a dangerous new era of nuclear proliferation has begun."[88]

First, let us trace this epidemic to its source. In an incisive piece, David Doyle, novelist and veteran of the OSS, CIA, and U.S. Foreign Service, observes that the conversion of the fifteen Soviet republics into the Commonwealth of Independent States (CIS) has created some worrisome situations. One problem is "the future of some 33,000 strategic and tactical nuclear weapons, and uncounted amounts of military nerve agents." Another problem is that "a few of the newly independent states, notably Armenia and Azerbaijan, are able and willing to use modern weapons against each other on a scale that can quickly become outright war."[89]

Doyle believes that whereas the prospects of CIS members doing battle with one another in a non-nuclear civil war are horrendous, resort to nuclear weaponry would be devastating for the whole world. Some 11,000 strategic (long range) missiles are temporarily controlled by CIS armed forces commander Marshal Yevgeny Shaposhnikov, and the firing codes are presumably in the hands of Russian President Boris Yeltsin. These weapons are mostly in Russia, with some in Ukraine, Kazhakhstan and perhaps Belorus. Hopefully, none of these states seems likely to try to use them before they are destroyed.

But the other two-thirds, some 22,000 tactical nuclear weapons constitute a different story. These weapons—the smallest as powerful as the bomb that obliterated Hiroshima—are spread over the former U.S.S.R.... Despite assurances from the CIS, revelations of theft of huge amounts of military assets by former Communist Party leaders (such as the Kola Society) for sale in the private sector are very unsettling. Also, hungry and disillusioned military weapons specialists are reportedly already filtering out of the CIS with their technology if not with actual hardware. There are a lot of willing buyers out there, ready to pay cash: North Korea, Libya, Iran, Iraq, Pakistan, India, to mention a few.[90]

In February of 1992, CIA Director Robert Gates told a congressional committee that the North Koreans were only a few months away from building an atomic bomb. This puts South Korea and Japan under a potential nuclear threat and jeopardizes the safety of 39,500 U.S. troops in South Korea. North Korea is ruled by Kim II Sung, a dictator who launched the Korean War in 1950, and who has practised international terrorism. To make the bomb, he has repeatedly had to deceive the International Atomic Energy Agency. With the clock ticking on North Korea's nuclear bomb, interested nations face a dilemma: "They can either stand by and watch an 80-year-old megalomaniac arm himself with nuclear weapons. Or they can take out Yongbyon and risk war with North Korea. It is a fateful choice, but it cannot be avoided."[91]

The Gulf War revealed that the West had seriously understimated the timetable for Saddam Hussein's atomic programme. The Iraqi Manhattan Project was very close to making Iraq a nuclear power. As terms of surrender, the United Nations had ordered the Iraqis to destroy their nuclear plant. Iraq has been denying that it was trying to produce nuclear weapons, but UN inspection officials have uncovered evidence of an extensive programme that was close to developing a warhead.

The cases of North Korea and Iraq illustrate two points of international gravity. First, in respect to making a nuclear bomb, things that were very difficult for great scientists in 1943 are now relatively easy. Second, the collapse of the economy of the former Soviet Union has unloaded supplies of uranium ore and other nuclear materials on to world markets, and inspectors warn that "it may be only a matter of time before more dangerous products, including tons of plutonium from spent Soviet reactor fuel and perhaps even uranium-processing technology from the Central Asian republics, reach the black market."[92] The stark economy is also driving Russian physicists and multiple warhead specialists to sell their expertise to Third World countries willing to pay several hundred times their former salaries.

Privately U.S. officials concede that their system to prevent the spread of nuclear weapons, based on a strategy of secrecy, export controls, and inspections of civilian nuclear reactors, has failed. Victory in the Gulf War presented an opportunity to make this passive system into an assertive nonproliferation system, but that opening was not seized. Consequently, a new and much more dangerous era of nuclear proliferation has now begun.

## REFERENCES


1. In "A Statement from President Nixon," *The Environment*, editors of *Fortune* (New York: Perennial Library, 1969), p. 11.
2. *Ibid*.
3. *Ibid*., p. 12.
4. Max Ways, "How to Think about the Environment," *Environment, Ibid*., p. 204.
5. "Reconciling Progress with the Quality of Life," *Ibid*., p. 8.
6. Ways, loc. cit.
7. L. White, "The Historic Roots of our Ecologic Crisis," in *Ecology and Religion in History*, ed. David and Eileen Spring (New York: Harper Torch Books, 1974), p. 23.

8. Sancton, *op. cit.*, p. 27.
9. John C. Ryan, "Conserving Biological Diversity," *State of the World, 1992*, ed. Lester R. Brown (New York: W.W. Norton1992), p. 9.
10. Eugene Linden, "The Death of Birth," *Time,* January, 2, 1989, p. 32.
11. Madhav Gadgil, "Saving the Subcontinent's Wealth," in *The Hindu Survey of the Environment 1991* (Madras: National Press, 1991), p. 140.
12. *Ibid.*
13. *Ibid.*, p. 141.
14. *Ibid.*
15. Ryan, *op.cit.*, p. 10.
16. *Ibid.*, pp. 10, 11.
17. Sancton, *op.cit.*, p. 34.
18. S.S. Chawla, "Punjab-Haryana: Like Woes of Neighbours," in Gadgil, *op.cit.*, p. 53.
19. Sunny Sebastian, "Rajasthan: Digging at Roots of Sanctuaries," in Gadgil, *op.cit.*, p. 53.
20. V.T. Joshi, "Madhya Pradesh: Not Learning from Experience," In Gadgil, *op.cit.*, p. 59.
21. M.S. Prabhakara, "Assam: Pubic Sector Playing Havoc, " in Gadgil, *op.cit.*, p. 63.
22. Anil Bhattacharya, Tripura: Gunti Dam, of Uprooted People," in Gadgil, *op.cit.*, p. 65.
23. P.K Roy, "Utter Pradesh: Heading for Ecodisaster" in Gadgil *op.cit.*, p. 56.
24. S.S. Chawla, "Himachal Pradesh: The Dying Lake Renuka," in Gadgil, *op.cit.*, p. 51.
25. M. Vannucci, "Mangroves: Saving a Crucial Ecosystem," in Gadgil, *op.cit.*, p. 157.
26. Ryan, *op.cit.*, p. 12.
27. *Ibid.*

28. *Ibid.*, p. 9.

29. Linden, *loc. cit.*

30. Ryan, *op.cit.*, p. 13.

31. *Ibid.*, p. 14.

32. Lonnelle Aikman, "Nature's Gifts to Medicine," *National Geographic,* September 1974, p. 424.

33. Ryan, *loc. cit.*

34. *Ibid.*

35. Michael D. Lemonick, "The Ozone Vanishes," *Time,* February 17, 1992, p. 63.

36. A.P. Mitra, "The Greenhouse Effect," in Gadgil, *op.cit.*, p. 77.

37. Lemonick, *op. cit.*, p. 60.

38. *Ibid.*

39. Michael D. Lemonick, "The Heat Is On," *Time,* October, 19, 1987, p. 59.

40. Peter Wagner, "Hawaii's Toxic Sins Pale in the Glare of U.S. Problems," *Honolulu Star-Bulletin,* August 17, 1989, pp. A1-2, B1-2.

41. Samuel S. Epstein, "Environmental Ethics, Health and Human Disease," *Encyclopaedia of Biothics,* ed. Warren To. Reich (New York: The Free Press, 1978), vol. 1, pp. 380-381.

42. *Ibid.*, p. 380.

43. *Ibid.*

44. *Ibid.*

45. The Harvard Medical School Health Letter Book, ed. G.T. Johnson and S.E. Goldfinger (New York: Warner Books, 1981), p. 367.

46. *Ibid.*, pp. 382, 383.

47. *Ibid.*, p. 382.

48. Ellen Hale, "Communism Is Blamed for Deadly Pollution

in Eastern Europe," *Honolulu Star-Bulletin,* August 14, 1990, p. D-11.

49. *Ibid.*

50. *Ibid.*

51. *Ibid.*

52. Glenn Garelik, "The Soviets Clean Up Their Act," *Time,* January 29, 1990.

53. *Ibid.*

54. Christine Gorman, "Mexico City's Menacing Air," *Time,* April 1, 1991.

55. Indranil Banerjie, "The Invisible Menace," *India Today,* June 30, 1984, p. 142.

56. *Ibid.*

57. *Ibid.*

58. *Ibid.*

59. *Ibid.*

60. B.B. Sundaresan, "Industrial Pollution: The Main Culprits," in Gadgil, *op.cit.,* p. 99.

61. Angus Denning and Kathy Cheng, "Victim of Its Own Success," *Newsweek,* June 4, 1990, pp. 74, 75.

62. *Ibid.*

63. *Ibid.*

64. Edward Flattau, "Japan Has a Split Environmental Personality," *Honolulu Star Bulletin,* March 19, 1991.

65. *Ibid.*

66. Eduard Shevardnadze, *The Future Belongs to Freedom* (New York: The Free Press, 1991), p. 171.

67. *Ibid.,* pp. 174-175.

68. *Ibid.,* pp. 175, 176.

69. "Nuclear Energy," *Honolulu Star-Bulletin,* July 15, 1990.

70. John Langone, "Waste: A Stinking Mess," *Time,* January 2, 1989, p. 44.

71. *Ibid.*

72. *Ibid.*

73. *Ibid.*

74. Dan Grossman and Seth Shulman, "Down in the Dumps," *Discover,* April 1990, p. 38.

75. George J. Church, "Garbage, Garbage Everywhere," *Time,* September 5, 1988.

76. *Ibid.*

77. Al Gore, *Earth in the Balance: Ecology and the Human Spirit* (Boston: Houghton Mifflin, 1992), p. 151.

78. Church, *Loc. cit.*

79. Gore, *op.cit.,* pp. 154, 155.

80. Eugene Linden, "the Last Drops," *Time,* August 20, 1990.

81. Eugene Linden, "The Last Drops," *Time,* August 20, 1990.

82. Lee-Jay Cho, "The Population Institute," *East-West Center Prospectuses* (Honolulu: 1982-83), p. 1.

83. Strobe Talbott, "How Bush Has Wimped Out," *Time,* December 16, 1991.

84. "UN Rings Alarm on Populations Rise," *Honolulu Star-Bulletin,* May 13, 1991.

85. Aldo Leopold, *A Sand County Almanac* (New York: Oxford University Press, 1968), p. 224f.

86. Anastasia Toufexis, "Too Many Mouths," *Time,* January 2, 1989, p. 49.

87. Talbott, *loc, cit.*

88. Stephen Budiansky, "the Nuclear Epidemic," *U.S. News and World Report,* March 16, 1992, p. 40.

89. David Doyle, "Intelligence Needs Heat Up as Cold War Ends," *Honolulu Star-Bulletin,* March 23, 1992.

90. *Ibid.*

91. B.J. Cutler, "Foreign Affairs," *Honolulu Star-Bulletin,* March 20, 1992, p. A-16.

92. Budiansky, *loc. cit.*

# 10

# HINDU ETHICS OF ENVIRONMENT

We examined the current environmental crisis with reference to the biosphere, global warming, waste, pollution, population, and nuclear proliferation. *These are our latest killing fields.* Our devastation of the environment is the result of our fundamental failure to be able to see the planet as "a single abode."

As a result, we have thrown ourselves all the way back to our primitive beginnings, in which the dominant problem on earth was coping with the environment. The only difference between our tribal beginnings and the situation today is that the problems now are of our own making. The killing fields represent our tragic failure to manage this planet. Through our capacity for invention and our sense of creative splendor, we have found little difficulty to make the earth *accessible*, but never *manageable*. Now the price of our survival is the management of our planet.[1]

This calls for a "philosophy of the whole" and for "a new education" in the comprehension and management of our planet.[2] It will be not so much a test of our skill but of our philosophy, for "everything begins—or ends—with a view of life."[3] Conservationist Aldo Leopold argues for the need of a land ethic as the next step toward productive ways of relating to our natural environment.

The first ethics dealt with the relationship between individuals. The Mosaic Decalogue is an example. Later accretions dealt with the relationship between the individual and society. Christianity tries to integrate the individual to society, Democracy to integrate social organisation to the individual.

There is yet no ethic dealing with man's relationship to land and to non-human animals and plants which grow upon it. Land, like Odysseus's slave-girls, is still property. The land relation is strictly economic, entailing privileges but not obligations.[4]

It is the contention of this study that the sort of land ethic Leopold is pleading for may be found in the ecological insights of Hinduism. Hindu philosophy is a rich repository of ideas, attitudes, and values which can furnish the necessary principles for a proper management of our physical environment. We propose (1) to sketch the development of these ideas; (2) to outline the ecological principles derived from them; (3) and then to see how these principles may be applied to the problems of the biosphere, global warming, waste, and so on.

The first glimmers of Indian environmental thinking are suggested by the excavations of the Indus Valley civilisation (3000=1500 B.C.). The pre-Aryan religion of this time is captured in figurines of a mother goddess and representations of a horned god accompanied by animals and plants. The prominence given to fertility suggests a religious orientation in which nature plays a central role.

This fecund religion of the Indus Valley was one source of the Hindu tradition as it took root in the Vedic age (1500-600 B.C.). A second source was the religion of the people who spoke the Dravidian languages and whose deities were also zoomorphic. Animals were revered so highly that they were considered fit to symbolise the gods. These precedents contributed to the making of Vedic religion into a terrestrial communion in which nature was alive! Embree explains:

The agricultural life of the people combined with the special features of the Indian climate—the brilliance of the sunshine, the clarity of the atmosphere, the easy beneficence quickly altering to harsh malevolence—made the people sensitive to the natural world. Much of the imagery of the poetry and religion can be understood as the response of the imagination to a Nature that seemed living and animate.[5]

An example of the poet's sensitivity to the natural world is the compendium of hymns to *Ushas,* godess of Dawn. They typify the resonance felt by the ancient Indians between the phenomenal world and human concerns. "Perhaps at this early stage can be seen the characteristic tendency of Indian literature to see nature as a frame from human emotions. The natural world is not understood as something apart from man, but as a reflection of his moods and passions."[6]

Like *Ushas*, Aranyani, tutelary goddess of the forest and wilderness, is celebrated and worshiped. The forest (*aranya)*) is perceived as different from the village *(grama)*. Life in the forest is seen as harmonious, with songs of many birds and creatures, cattle grazing, wagons carrying heavy loads of harvest, and wood choppers felling trees. the "Forest Queen" is praised as "sweet-scented, redolent of balm." she is "Mother of all sylvan things, who tills not but hath stores of food." The picture is clearly one of balance and prosperity. "Man eats of savory fruit and then takes, even as he wills, his rest," But through this "Lady of the Wood" is an object of familiar beauty, the element of mystery dominates. Part of the mystery is Aranyani's capacity to strike back. "The Goddess never slays, unless some murderous enemy approach."[7]

The heart of Vedic worship was rituals intended to adapt the beneficent powers of nature to human needs. The gods to whom the prayers were addressed granted favours but were, in turn, dependent on humans for offerings in *yajna.* The picture is one of interrelationship and interdependence between people and the great cosmic forces that were deified. The gods also helped one another and often worked in pairs, indicating the systemic integration of natural forces.[8]

This highlights a significant feature of vedic religion: the notion of cosmic law. Philosophers were not content to study human nature only, but turned their speculations on the larger world of nature to which humans belonged. They viewed the external world as an ordered whole, presided over by nature gods. Indeed the gods are invested to account for the cosmic character of the universe.

The ideas of order and unity are strengthened as the religion evolves toward monotheism and monism. The notion of physical order is combined with a belief in ethical order, and both are brought together in the concept of *rita.* It was early recognised that there were certain proper ways of relating, both to people and to animate and inanimate objects, and that these forms of relating merited rewards and punishments. Consequences counted. Cosmogonic speculations entertained theories of creation and evolution. The early notion that the universe was created by the gods was superseded by the effort to find origins in some primary principle.

Among the early cosmogonic hymns, the hymn to Goddess Earth invites attention. The earth is lauded as a mistress: "The golden-breasted resting-place of all living creatures." No discrimination is made between humans and other species: "The mortals born of the live on thee, thou supportest both bipeds and quadrupeds."[9] There is also an awareness of conservation: "What, O earth, I dig out of thee, quickly shall that grow again: may I not, O pure one, pierce thy vital spot, (and) not thy heart!"[10] Most profoundly, the Hymn to the Earth *(Prithivi)* expresses the conviction that the land which sustains and supports us is the embodiment of the divine Mother:

O Earth, thy centre and thy navel, all forces that have issued from thy body—

Set us amid those forces; breathe upon us. I am the son of Earth, Earth is my Mother.[11]

The summit of the Vedic age is reached in the teachings of the Upanishads. Here, the highest ideal is not living in harmony with the will of the gods or sacrificial rectitude, but self-knowledge. Vedic ceremonialism and caste duties give way to a new spiritual quest which turns inward for the purpose of understanding Ultimate Reality.

The motivation behind this introspective search is the fundamental idea running through the early Upanishads that "underlying the exterior world of change there is an unchangeable reality which is identical with that which underlies the essence of man."[12] "Verily, he is the great, unborn Soul, who is this [person] consisting of knowledge among the senses. In the space within the heart lies the ruler of all, the lord of all, the king of all."[13]

The formulation of this ideal was a long time in the making. Summarily stated, the Upanisadic sages synthesised two notions which originally had totally different connotations, namely, *Brahman,* or the ultimate source of the external world, and *atman,* or the inner self of the person. These sages teach that the *Brahman* of the macrocosm is none other than the *atman* of the microcosm—"That self is, indeed Brahman."[14] With *Brahman* equated with *atman,* each term came to signify the eternal ground of the universe, including the ground of the being of humankind and the being of nature.

The Upanishads also witnessed the development of the doctrine of *karma*, which, as the law of cause and effect, early attracted the Indian mind. The Vedic hymns are connected by the conviction that one thing we can be certain of in this world is that whatever we do must have consequences. *Rita,* or order, is an inescapable fact of life, which is binding on mortals and gods alike. Herein are the germs of the doctrine of *karma*, which, by the time the earliest Upanishads were written, is combined with the idea of rebirths. It is held that the universe is so structured that the consequences of our actions are inexorably played out into an unknown and unseen future.

The Upanishads record diverse theories of creation. The most succinct formulation is found in the Isha Upanishad. Recognising the unity that underlies the diversity of the world, the sage declares: "By the Lord (*isha*) enveloped must this all be—Whatever moving thing there is in the moving world."[15]

The epic and Puranic literature contain many imaginative theories of creation. In none of these stories are humans exalted over other creatures. To the contrary, to be created human invests people, not with rights and privileges, but with responsibilities and obligations.

Paralleling the Puranas are the Tantras, the scriptures of the Shaktas. From the perspective of environmental thinking, it is interesting to note that the Mother Goddess motif we had first encountered in the pre-Aryan religion is here (A.D. 300-750) combined with Aryan elements, giving prominence to the worship of Sakti, the Divine

Power personified as a goddess. In the form of Durga, the goddess was identified with the Absolute of Vedanta philosophy.

Hindu environmental thinking reaches its apex in the philosophical sutras of the six schools called *Darshanas,* or views of life. We limit our remarks to the ecological insights of the two best-known schools: Samkhya and Vedanta.

The Samkhya system is ontologically dualistic. It postulates two uncreated, independent, and eternal principles known as *purusha* and *prakriti. Purusha,* the spiritual principle, and *prakriti,* the material principle, are the transcendental essences of our conscious and unconscious experiences. All experiences arise from this duality between *purusha,* the knower, and *prakriti,* the known. Together they form the matrix from which the phenomenal universe evolves.

*Prakriti* (nature), the changing object, is the *causa materialis* of the whole world of becoming. Samkhya affirms the reality of matter. The theory of causality explaining this evolution is known as *parinamavada.* It holds that the effect is only a modification of the cause. The effect resides in the cause in a potential form prior to manifestation.

The existence of *prakriti* is proved through inferential reasoning, and not by recourse to revelation The nature of *prakriti* is also explained by reason. Since, the effect is only a modification of the cause, the nature of the cause can be deduced from the nature of the effect. Upon analysis, the physical universe is seen to possess three basic properties called *gunas (sattva, rajas, tamas)*; therefore, it is concluded that the Unmanifest cause must also be characterised by similar constituents.

When the *gunas* are in a state of equilibrium, there is

no evolution. The *gunas* maintain their dynamic character, but do not interact with one another.

Change within unconscious *prakriti* from the state of dissolution (*pralaya*) to that of evolution (*sarga*) is attributed to the proximity of conscious *purusha (Sannidhi matra*). Acting like a magnet, the presence of *purusha* draws the *gunas* into a state of excitation, and so starts them off on their evolutionary cycle. By combining in various proportions, the *gunas* give rise to the manifold variety in the physical world. The union between conscious though inactive *purusha* with active though unconscious *prakriti* is compared to the union that exists between a blind man carrying a lame man on his shoulders. "From its union proceeds evolution."[16]

What is environmentally noteworthy in Samkhya's theory of evolution is that, though the universe is viewed dualistically, spirit and matter are held together in a teleological balance. Further, unlike some systems in which spirit and matter are polarised as good and evil, the Samkhya system suggests no such moral dualism. To the contrary, both the design and function of *prakriti* are aimed at the liberation of *purusha.* The *Karika* says: "As the insentient milk flows out for the growth of the calf, so does Nature act towards the emancipation of spirit."[17] Hiriyanna comments: The noteworthy point here is the physical accompaniment of man as well as his environment is neither hostile or indifferent to his attaining the ideal of freedom. Through them rather, *Prakriti* is ever educating him into a fuller knowledge of himself with a view to securing that result. Nature therefore, cannot in the end, be said to enslave spirit. In fact, it behaves towards man as a "veritable fairy godmother."[18]

The second of the *darshanas* to which we turn is the Vedanta system. The chief commentators are Shamkara (eighth century A.D.) and Ramanuja (twelfth century

A.D.). They provide us with nontheistic and theistic interpretations of the Vedanta, respectively. Shamkara's system is monistic, and is known as *Advaita.* Ramanuja follows a qualified monism called *Vishishtadvaita.*

The purpose of Advaita is the intuitive realisation that there is one and only one Reality which is the negation of all plurality and difference. This rigorous monism is the key to Shamkara's system. It is capsulised in the well-known phrase: *Brahma satyam jagan mithya jivo brahmaiva na parah* ("*Brahman* is real, the world is false, the self is not-different from *Brahman*").

The environmental aspect of this affirmation is that *Brahman* is the sole Reality. It is the ground of all existence, and individuals in their essential being are identical with *Brahman.* To say that "the world is false" does not mean that the Advaitan thinks the world a phantom, something imaginary or nonexisting. The Advaitan is not a subjective Idealist, believing that there is no exramental reality and that all we see and know are only ideas in our mind which are falsely thought to be outside the mind. To the contrary, the world as we see it does indeed exist outside our consciousness and has positive content. Yet, it is unreal. How?

For the Advaitan, the test of truth consists in what might be called "unsublatedness." A thing is untrue or unreal when it is sublated by a subsequent experience. That alone is true or real which is unsublated or noncontradicted (*abadhita*) by subsequent experience.

Judged by this test of truth, the world is both real (*sat*) and unreal (*asat*). It is real because it has a positive, concrete existence, and to that extent is not an hallucination. It is unreal because it does not meet the test of Reality, being impermanent and subject to constant alterations. The ecological meaning of this philosophic reasoning is that the truth of Reality lies in its pervasive unity, and therefore to think of the world as separate or independent entity is to indulge in an illusion.

Next we look at Ramanuja's qualified nondualism. According to Ramanuja, *Brahman* is not a characterless identity; it possessed features and qualities. He agrees with Shamkara that *Brahman* has nothing different from it, of either a like kind or unlike kind. But he disagrees with Shamkara that *Brahman* is without internal differences. He asserts the *Brahman* is the sole Reality. There is nothing besides *Brahman.* But *Brahman* is a determinate being, having internal differences, and it maintains its identity in and through the differences internal to it.

Thus *Brahman* is an organism which sustains, as parts integral to it, the world of matter *(acit)* and the world of souls (*cit*). These two orders are internal to *Brahman,* and though dependent on *Brahman* and grounded in it, cannot ever be resolved away into an indistinguishable identity with it.

Ramanuja uses various analogies to describe this total integrity called *Brahman.* One is the relation between the soul and the body it animates. *Brahman* is the soul, and the world of matter and the world of individual souls constitute the body which the soul animates. The body is indeed distinct from the soul, and though it is distinct it has only a dependent and subordinate existence because the body cannot survive without the soul.

Another example is the relation between substance and qualities. Qualities are distinct from the substance, and yet they have only a dependent existence and can never have a self-subsistent existence. They are totally dependent on the substance. Substance and quality together constitute a unity—not a simple and bare unity, but the unity of a complex whole. Brahman, the world, and individuals constitute such a whole.

This brief survey has shown that Hindus have reflected long and hard on the relation of God to humans, God to physical nature, and people to the world. These assumptions serve as the backdrop for formulating ethical principles that are both philosophically valid and environmentally sound. First, we shall articulate what these principles are in the context of the current environmental debate; and we shall then attempt to apply them to actual situations.

## THE PRINCIPLE OF UNITY

Indian speculative wisdom is essentially unitive and cosmic. The Upanishadic sages behold the Supreme Being as manifest in all creation. "Verily, the Soul is the overlord of all things, the king of all things. As all the spokes are held together in the hub and felly of a wheel, just so in this Soul of all things, all gods, all worlds, all breathing things, all selves are held together."[19]

The cosmic manifestation of the Supreme Being is expressed in evolutionary terms. Evolution is progressive and spiritual. It starts with matter in which the Spirit lies dormant. From a lifeless stone, Spirit rises to the level of life and manifests itself in the vegetable kingdom. It then ascends to the level of consciousness, manifesting itself in the animal kingdom. From consciousness, it progresses to the level of intelligence, and takes form in the kingdom of humans. The most perfect state is that of pure Spirit, in which matter lies dormant and Spirit manifests itself as pure bliss in the Supreme Spirit.

Thus, all the world is a stage, and each life form is a unique player. They have their exits and their entrances, and the One Spirit plays many parts. Matter is in tension with Spirit, but Spirit, step by evolutionary step, comes into its own. The Bhagavadgita summarises this cosmic scenario thus: "Unmanifest is the origin of beings, manifest their mid-most stage, and unmanifest again their end."[20]

That which precedes and succeeds the evolutionary process is shrouded in mystery because it is outside the pale of time. As mortals in this planetary theater,we cannot apprehend how the Divine Actors divided itself into subject and object and commenced this evolutionary show. Nor can we tell how Sirit shall prevail over matter so that the original perfection is finally restored. All that we know at this stage of history is that there is movement, that this movement is progressive and spiritual, and that the divine plan for human life is that we join the creative ascent of the Spirit. To have this knowledge is to be able to distinguish the relative merits of material, biological, intellectual, and spiritual values.

Such is the unitive vision of Hinduism. It is rationally held, but it derives its strength from subliminal springs. The conviction is religious, but it permeates the whole culture, including all the arts and sciences.

The cosmic vision of Hinduism sees humans as an intrinsic part of nature. The fabric of life has many special strands, and human beings are special in their own rights by virtue of their moral character; but this does not place them outside nature or above nature. *Distinction,* not *separation,* is the hallmark of the Hindu view of the positon of humans in nature. The unitive thinking of Hinduism aligns it with a "deep ecology" which places people *in nature*, as oppossed to a "shallow ecology", which is anthropocentric and which ascribes to *Homo sapiens* a position of dominance and superiority *over nature.*

The unitive thinking of Hinduism is also attuned to the holographic model, and to the new paradigm of science in which there is a radical shift from the parts to the whole. As described by Fritjof Capra, "In the old paradigm it was believed that in any complex system, the dynamics of the whole could be understood from the properties of the parts. In the new paradigm the relationship between the parts and the whole is reversed. The properties of the parts can be understood only from the dynamics of the whole. Ultimately, there are no parts at all. What we call a part is merely a pattern in an inseparable web of relationships."[21]

In the context of the new physics, the Hindu paradigm is clearly holistic, emphasising the whole over the parts. Yet, it would be more precise to describe Hindu ethics as being *ecological* rather than holistic. The distinction is fine, but important. Holistic thinking sees an entity in all of its complex relationships. For example, holistic medicine is sensitive to the body-mind relationship, and to how all parts of the human organism function together. Ecological thinking is also holistic, but it expands the unity to encompass larger systems which are all dynamically related. In Hindu ecological thinking, the microuniverse is one with the macrouniverse. The individual is a child both of earth and of the distant stars.

The philosophic contribution of the unitive world view of Hindu ethics is that it avoids the dichotomies that have characterised Western religious, philosophic, and scientific thinking—dichotomies between God and man, between God and the world, and between people and nature. This split-thinking is now scientifically discredited as being completely dysfunctional in the task of tackling environmental problems.

More positively, the unitive, evolutionary view of Hindu environmental ethics engenders respect and reverence as the *natural* way of looking at things in the world. They are all parts, but none is *apart,* for each is a part of the whole, and just as a spark of fire is the fire, so also each item of nature is instinct with spiritual life.

This awareness of biotic and spiritual connectedness

has engendered in the Hindu mind and attitude of noninjury to all living ceatures, as expressed in the doctrine of *ahimsa*. It is one thing to know of the continuity of humans and physical nature in terms of the biological evolutionism taught by Darwin, and quite something else to be inwardly aware of the fundamental unity of all life. The latter affects the whole person and mobilises the individual to act responsibly toward the weaker links in the chain of life.

Eliot Deutsch argues that if the unitive thinking of Vedanta were fully assimilated, it "would rule out the possibility of any gross exploitation of nature and of ourselves." It is certainly one of the greatest follies of man that he still believes that, as an individual (whether in corporated or not), he has an inalienable right to exploit and destry nature for his own personal gain, or indeed for the mere satisfaction of his own idiosyncratic taste. But surely nature does not *belong* to anyone. The idea of *possessing* or *owning* nature is utterly alien to Vedanta, as it ought to be to any sophisticated spiritual perspective. "Possession" has no meaning where there is no real possessor. If everything in nature is already one's own basic self, and the self of all others, the clearly no single individual or group of individuals can lay claim to it. God, if he exists, is certainly some kind of socialist. And if the Hindu gods laugh, as assuredly they do, then the skies over India must rock from time to time with hilarity over some men's nonsensical claims to ownership of the universe. The Hindu god Shiva, the god of destruction, is certainly quickening the steps of this cosmic dance, pleased that man has been so cooperative with him in his work of destruction.[22]

## THE PRINCIPLE OF INTERCONNECTEDNESS

The unitive orientation of the Indian mind has predisposed it to account for natural and human processes in terms of the principle of interconnectedness. The concept of *rita* representing the "immanent dynamic order" was an early formulation of this principle that linked cause and effect. Order was not something imposed by the gods from outside, but was the internal function of the universe to which all beings, including the gods, were subject. Later, *rita* gave rise to the notion of *dharma,* meaning "what holds together," and thus the basis of all order. The operative element of *dharma* is the law of *karma,* which, as far back as 1000 B.C., was affirmed thus by Indian *rishis:* "According as one acts, according as one conducts himself, so does he become."[23]

This notion of causality has given to Indian thinking a consequentialist bent which evaluates things in terms of their connectedness. Deutsch deftly brings out the relevance of the doctrine of *karma* for environmental planning. According the doctrine of *karman,* every deed or act that one performs has its effect in the world and forms a "tendency" *(samskara or vasana)* within the doer to act in a similar away in the future; to form, in other words, a *habit* of action. This habit of action produces further consequences in the actor and in the world and, accordingly, the wheel of life turns around and around. Whatever one does will have effects not only in the immediate present but in the long future as well: any act, in short, will have consequences that reach out far beyond the act itself. And everything in nature is so interconnected through causal chains and relationships that we ourselves become part of the natural process and are conditioned by it.[24]

The environmental crisis is basically made up of two factors; one, our belief that we are somehow separate from nature, referred to above, and two, our willingness to ignore the consequences of our actions. The latter problem proceeds from our myopic vision that concentrates on current needs and short-term issues. With little faith in the future, we find it difficult to assess the impact of our actions on future generations, and act as if the future is now.

One example is the DDT poisoning, notwithstanding the warnings of pesticide abuse by such brave pioneers as Rachel Carson in her classic book, *Silent Spring.* For a time, government official refused to look beyond short-term profits for long-term consequences.

Another example of foreshortened thinking was the use of the herbicide Agent Orange in the Vietnam War. The military focus was solely concentrated on the ability of this poison to obliterate the lush ground cover, making it easier to shoot the enemy out in the open. The effects of this herbicide on the soldiers, causing chromosomal damage and producing cancer, were not given a second thought. Particularly painful was the tragedy for the family of Admiral Elmo Zumwalt, who authorised the use of Agent Orange, because it made casualties of his veteran son and grandson. Wary of the fact that all chemicals have extraordinarily powerful effects on the world around us, Al Gore asks: How can we be sure that a chemical has only those powers we desire and not others we don't? Are we really taking enough time to discover their long-term effects? Agent Orange, after all, is just one of the better-known examples of a whole new generation of powerful compounds created in the chemical revolution, which picked up speed after world War II; over the past fifty years, herbicides, pesticides, fungicides, chlorofluorocarbons (CFCs), and thousands of other compounds have come streaming out of our laboratories and chemical plants faster than we can possibly keep track of them. All of them are supposed to improve our lives, and hundreds of them have done so. But too many have left a legacy of poison that we will be coming to terms with for many generations. [25]

In addition to the consequential side of interconnectedness, there is the causal side. The ecological standpoint encompasses a view of the whole, and comprehends how the diverse parts of nature interact with one another, reaching toward a balance that is sustained through time.

The principle of interconnectedness demands that we should not compartmentalise the planet from the civilisation that exists upon it. Civilisation is part of that planetary whole, and when we view the whole we are viewing ourselves as constitutive of the whole. The sun and the sea are among the natural forces which interact with one another, but people, no less than Pinatubo, are a fore of nature, and if this feature of human causality is not assessed, then we will not be able to see how the pressures of civilisation are throwing the planet off balance.

## THE PRINCIPLE OF INTERDEPENDENCE

If reality is one, if all parts of the whole are interconnected, then interdependence must define the mode of ecological relationships.

Hinduism's earliest expression of interdependence takes shape within the dynamics of an awesome sense of all-encompassing natural forces, and is ritualised in the magico-religious form of *yajna*. Through sacrifice the worshipper attempted to secure earthly blessings because it was believed that the generative properties of nature were maintained by the gods, who in turn depended on humans to sustain them through oblations. Offerings to these divinised forms of natural phenomena acknowledged that to ensure fertility, humans must harmonise their lives with the energy of the sun, the wind, water, and fire. Delicate exercises had to be observed to make sure that the mechanisms did not go wrong.

In time, Hinduism outgrew the magical aspects of

*yagna,* but its sense of interdependence in the world at large persisted when the gods were seen as multiple manifestations of the One Source of creative energy, diversified in myriad life forms.

The most inclusive declaration of interdependence is found in the notion of *rina.* The *rinas* represent the congenital debts which each individual owes: to the sages, to the ancestors, and to the gods. These obligations are satisfied by study, by begetting a family, and by offering sacrifices. The underlying ethical principle is to give back what one has received. Through reciprocity, balance is maintained and welfare is assured. Manu warns that the person who seeks liberation without fulfilling his obligations incurs hurtful consequences (*vrajatyadhah*).[26]

We *propose* that the traditional notion of *rina-traya* be made more inclusive by expanding the Hindu's obligations to society to include *obligations to the land.* We may designate this as the fourth debt or *caturtha-rina,* and call it *bhu-rina.* It should stand for our obligations to the *Bhumi,* mankind's terrestrial home. This proposal is not as innovative as it appears because the elements essential for reciprocal land relationships are vitally embodied in the ancient theme of the earth as Mother Goddess.

*Prithivi,* and Goddess Earth is generally invoked with heaven (Dyaus), the parents of the universe. Maurer observes that the concept of *Prithivi* in the *Rigveda* (5.84) "is much broader than ours might be of an Earth goddess, as it includes also the atmospher or area of the rainclouds just above the earth. *Prithivi* in this extended sense...is looked upon as a great animator of all things as well as the physical support of the mountain and the trees of her surface."[27]

In her role as Mother, *Prithivi* is described as *Agni's* womb which receives the embryo of existences. Prithvi vivifies the earth with her might, but she also undergoes pain when there are too many mouths to feed, extracting from her more than she can supply. And so she cries: "I cannot endure all these people," and besceches Brahma to reduce the population. Brahma creates Death, who, in the form of a beautiful woman, sheds tears of sorrow which become fatal diseases.[28]

The metaphor of Earth as divine Mother is aptly chosen in the Hindu tradition because it fosters ties of kinship and evokes sentiments of care and protection. Infact, there is something elemental about this motif because it reaches back to the Indo-European peoples, and further, to mankind in general.[29]

But beyond the stuff of myths ecologists have scientifically traced "the complex reciprocities and mutualities among organisms, the food chains linking diverse species, the interlocking cycles of elements and compounds, and the delicate balances that are easily upset."[30] The very concept of the "ecosystem" carries the meaning of the interdependence of all forms of life in biotic communities, and underscores the long-range damage that can be incurred by thoughtless human intervention. Al Gore: In distinguishing between what is still uncertain and what is known about this crisis, it is important to emphasise that one thing we know quite well is that nature exhibits a recurring pattern of interdependency among the parts of the ecological system. We should assume with great confidence that if we upset the ecological balance of the entire earth in one way, we will upset it in related ways as well. Consequnetly, while a given behaviour may at first seem harmless in that part of the environment we are able to observe, we are unlikely to know enough about the effects of what we are doing to predict the consequences for other parts of the system—precisely because all its parts exist in a delicate balance of interdependency. [31]

## THE PRINCIPLE OF RESTRAINT

Ever since the Industrial Revolution, Western societies have made of *pleonexia* a commonplace. This has been especially true of the United States, which is the world's biggest user of natural resources and a major despoiler of the global environment. Because of the magnitude of its economy, the U.S. consumes one fourth of the world's energy each year. Yet, for a given amount of energy, it produces less than half as much economic output as Japan or Germany. Therefore, the issue is not just one of size but of greed and waste.

During the oil embargo of the early 1970s and 1980s, when energy was scarce and expensive, Americans tightened their belts and were economically more productive. Between 1973 to1985, our per capita energy consumption fell 12 per cent, and the average amount of goods and services per person rose 17 per cent. However, since those lean years, energy prices have been reduced, with a corresponding rise in energy consumption. "Americans, who own more than 135 million cars, or about one-third of the world's total, have been driving more and have resumed their love affair with large gas-guzzling cars."[32]

Such prodigal ways cannot go on indefinitely because our resources are not limitless, and when those that are non-renewable are consumed or transformed, they can never the replenished. The specter of "Old Mother Hubbard" requires that we redefine economic growth. Time-honoured principles of industry, such as economic growth based on increasing expansion and consumption, must now be questioned and corrected.

The problem is not only economic but moral. No nation has the moral right to overutilise the world's resources for its own selfish interests, without thought for future generations and the survival needs of poor people.

Emphasising the need for a conservation ethic, Kenneth Boulding held that it is "more likely to be found in Eastern religions than in the Judeo-Christian tradition," and therefore he urged Western societies to enter a genuine dialogue with Asian religions. It may be that India, for instance, had to adjust to a highly conservationist economy before the West had to do it, simply because the West has always been more expansionist geographically and hence should preserve the illusion of the illimitable plane for a longer period. There is a paradox here, in that the scientific revolution could probably have taken place only in the West, precisely because of the western expansionism and aggressiveness and the conept of man as the conqueror of nature. The East has never had any illusions about being able to conquer nature, and has always regarded man as living in a somewhat precarious position, as a guest of doubtful welcome, shall we say, in the great household of the natural world. [33]

Actually, India's conservationist attitude is less a matter of exigency as of ethics. The principle of restraint was early built into the Hindu mind's enchantment with renunciation. The hero in Hinduism is not the warrior but the ascetic. This is not to say that the pursuit of wealth is discouraged, but that it is always subject to the concerns of morality. *Artha* is relative to *dharma.*

Hindu ethics therefore does not reject growth, for that would consign the Third World to continued poverty and disease. Rather, it steers a path between "pro-growth" and "no-growth" and advocates a position of selective growth based on equity and quality.

In the name of global justice, the principle of restraint requires that the economic growth of advanced nations be directed toward services that are not resource intensive; and in the name of ecological wisdom the principle of restraint requires that wealthy nations limit their consumption and use their vast resources for research and development to produce technologies for recycling and waste reduction.

Having identified the environmental principles of Hinduism, our next task is to see how these ideas of *unity, interconnectedness, interdependence, and restraint* are able to address the environmental problems of biodiversity, global warming, waste, population, and nuclear proliferation.

## Biodiversity

On biodiversity, the problem is that people are recklessly exterminating life on earth. It has been predicted that over the next thirty years, humans will force one hundred species into extinction every day. Given its evolutionary outlook, Hinduism accepts extinction as part of life, but such vast killings are *himsa* in its gravest form, for the wiping out of whole ecosystems—nature's laboratories of new life forms—constitutes the "death of birth."

Biodiversity has caught the public attention by featuring high profile animals such as elephants, tigers and rhinos. But Hinduism's nondiscriminatory view of all life forms also calls attention to the plight of the *lowliest plants and insects.* When these species disappear unnoticed, "they take with them hard-won lessons of survival encoded in their genes over millions of years." [34]

Where the war on plants and animals is spurred by simple greed, Hindu morality condemns it as unworthy of human beings. Where the destruction is caused by poverty and population, as developing nations struggle to feed their burgeoning numbers and raise cash to make payments on international debts, Hindu ethics is sensitive to the moral and practical difficulties of their situation. On the one hand, wealthy developed nations ask the debtor nations to deliver on their interest-laden payments, and on the other hand, they want their clients to preserve, for the welfare of the world, the very forests which they chop down to finance those debts. For many Third World countries, it is a crisis of survival: they do not have the luxury of being concerned about global biodiversity when they must first feed their children.

Compassion requires that the rich nations reduce the debt burden of the poor countries. This plea was heard repeatedly at the Earth Summit, where poverty was described as the world's "quiet emergency." I was deeply felt that the most difficult aspect of moving toward sustainable development was the divide between North and South. Calls were heard for the North to accept big debt write-off in the South.

At the same time, the poor countries must become aware of the economic consequences of their short-term expedients, such as deforestation. In the long run, this is bound to backfire when their precious resources are depleted and the life forms they support become extinct. They then kill the goose that lays the golden egg. Karmically stated, as quick economic return often has a long fuse.

In order to prevent this from happening, environmental reviews should be conducted at the time when loans are being negotiated in the effort to prevent local banks from supporting projects that wipe out species. Also, governments, businesses, and the general populace need to be educated in the value of genetic diversity, and be shown that when habitats are destroyed the loss of the species living in those habitats is irreversible.

On the constructive side, 'gene banks' should be set up in places such as zoos too perpetuate species; and projects should be created to validate the fact that,with judicious planning, endangered habitats can be developed into profitable enterprises, without necessarily incurring their depletion or demise.

Further, the myth must be laid to rest that the preservation of biodiversity is intrinsically opposed to economic returns. This misperception was articulated by Secretary of State James Baker at a forum in Princeton, New Jersey. Defending the decision of the Bush administration not to sign a treaty at the Earth Summit intended to preserve the diversity of plant and animal life, and citing controversial provisions about patents, advanced biological genetic engineering and funding as the reasons, Baker declared, "This so called biodiversity treaty . . . would put people out of business in this country."[35]

The position of Hindu ethics has always been: *artha* and *dharma* can coexist. The experience of the ancient Indian *vaidya* who cultivated exotic plants for the development of a profitable pharmacopoeia demonstrates that there are dividends from diversity. The following are modern examples of the harvesting of unexpected benefits from exotic plants and animals:

* Squibb used the venom of the Brazilian pit viper to develop Capoten, a drug for high blood pressure.
* By transplanting genes from tropical tomatoes, the NPI biotech firm increased the density of U.S. tomatoes 2 per cent, promising catsup manufacturers extra profits.
* Scientists believe that arcelin, a natural protein in wild Mexican beans that repels insects, might protect somer U.S. crops without poisoning soil and water.
* Future newspapers may be printed on paper from kenaf, and African plant that can produce five times as much pulp an acre than the trees normally cut for newsprint.[36]

In addition to such facts, the position that the preservation of biodiversity can only be purchased at the expense of jobs is at odds with developments in Japan and Germany. These countries are recognising that a new world order is being born and that in this green revolution there will be an increasing demand for environmentally friendly products, and they are technologically preparing to become leaders in this global market. If this is correct, then the pitting of jobs against the preservation of biodiversity is a false choice. It is antithetical to a major way in which the world is today beginning to think, and tomorrow will begin to buy.

An alarming feature of the Bush administration was the operation of the White House Council on Competitiveness, headed by Vice President Dan Quayle. It had the power to grant exceptions to the Endangered Species Act, and in this capacity was dubbed the "God Squad." In May 1992, over the protests of the administrator of Americans Environmental Protection Agency, William Reilly, Quayle's Committee gave loggers permission to cut down 688 hectares (1,700 acres) of ancient forest in the Pacific Northwest. Asked by reporters about the "God Squad," Reilly responded, with a touch of bitterness, that it was "a group of people, of which I am a minor divinity, which has the power to blow away a species."[37]

There can be no room for a "God Squad" within the Hindu Pantheon. The Vaishnava tradition connects the evolution of animal life forms with divine incarnations, moving upward from fish, to amphibians, to mammal, and to humans. Karan Singh points out: "This view clearly holds that man did not spring fully formed to dominate the lesser life forms, but rather evolved out of these forms himself, and is, therefore, integrally linked to the whole of creations."[38]

If indeed there were room for a "God Squad" in the Hindu fold, the strong ones would be obligated to protect the weak ones. Gandhi incorporated this value into his teachings on cow protection. He thought cow protection was "the gift of Hinduism to the world." It symbolised the protection of the weak by the strong. It subsumed everything that feels. Causing pain even to the lowliest creatures on earth was a breach of the principle of "cow protection."[39]

Incidentally, "cow protection" in Hinduism should be seen as illustrative of the combination of utility and sanctity. All animals are sacred, but the cow is held specially so. It is the darling of Lord Krishna, and is a goddess who embodies several divinities. Yet, all of this veneration proceeds from the utilitarian role of the cow within an agrarian economy.

## Global Warming

The second problem Hindu ethics must confront is global warming. The danger looming over our planet is that greenhouse gases could bring about a climatic calamity. The dust from Mount Pinatubo has had the effect of global cooling, but this temporary phenomenon is not expected to exceed 1° F for the next few years, after which the greenhouse warming will again pick up.

Global warming was the second area in which Washington took the heat at the Earth Summit. The Bush administration was roundly criticised because its negotiators succeeded in softening a major treaty to control the emission of carbon dioxide and other gases said by some scientists to cause global warming. Here, too, the reasons given by Washington for its go-slow policy were the costs involved in modernising business to cut down on gases, the lack of scientific evidence that global warming is now taking place, and the hazards involved in making future projections.[40]

Our argument is that if the levels of greenhouse gases are on the rise, as the 1991 National Academy of Sciences' report attests, then the problem needs addressing. The academy said: "In the absence of greater human effort to the contrary, greenhouse gas concentrations equivalent to a doubling of the pre-industrial level of $CO_2$ will occur by the middle of the next century.[41]

In keeping with the conservationist element of Hindu ethics, nations should promote energy conservation by implementing energy efficiencies; limiting warming gases at 1990 levels by the year 2000; reducing dependency on coal, oil, and gas; utilising renewable resources; taxing carbon dioxide emissions in excess of set standards; implementing alternate energy sources such as solar power; harnessing methane gas expelled from landfills and cattle feedlots; banning the production of CFCs used in the manufacture of plastic foam and as coolants in refrigerators and air conditioners; and in general, greening the energy system.

Certainly, the least expensive and most benign step to take in controlling greenhouse emissions and slowing global warming is tree planting. This could be done as a companion strategy to the more complex task of deforestation. Trees absorb carbon dioxide from the air and pump moisture into the air by evapotranspiration. Trees are the most effective producers of clouds. Proper placement of trees in cities could reduce energy consumption by 15 to 35 per cent. Trees also create "white noise" (rustling noise in wind) which helps block out the more shrill urban noises.

Tree planting can be undertaken by all segments of society, from individuals growing fruit trees in their backyards and municipalities planting trees along city streets, to the governments reforesting hills and valleys with the help of school children or factory workers.

The rationale for tree planting can be found at all levels, but its roots strike deepest in religious soil. Eastern and Western religions attribute sanctity to trees. They represent life, and embody the meeting place of the spiritual and physical. The ideological and cultural roots of reforestation are firmly embedded in Indian soil and provide an optimum rationale.

The Hindu tradition venerates flora with the similar blend of utility and sanctity it expresses for fauna, and both flora and fauna, along with humans are seen as existing in a state of mutual connections and dependent cites. The tradition goes back to the Indus Valley civilisation, where the pipal tree is viewed as representing the Mother Goddess. Later it is associated with god Vishnu. The banyan tree also stands tall and majestic—Mother Nature's own cathedral. Under its vaulted canopy, Lord Buddha received his enlightenment. The record has it that in his quest for illumination Gautama withdrew to Uruvela, "a pleasant spot and beautiful forest." The Buddha was acting on the Indian belief that nature speaks with many tongues to the deep levels of the human soul, and hence the practice of locating temples, schools and monasteries in arboreal settings. So, too, the forest was deemed the appropriate environment for the *Vanaprastha* (forest recluse) and for the *Parivrajaka* (ascetic). Trees provided these aspirants with not only food for the body but sustenance for the soul. Their pristine pillars formed living temples that soothed the senses and brought heaven down to earth. In such all-embracing environments, there was felt continuity between matter and spirit. Boundaries were as indistinguishable as the inter wining of the roots of the trees under the surface. It, therefore, only took a small flight of the imagination for the Shatapatha Brahmana to derive the strength, aroma, juices, and colour of certain trees from diverse parts of the body of none other than Prajapati.

In gratitude to the Bodhi-tree under which he received his enlightenment, the Buddha bade all his monks not to harm trees because they, too, possessed sensate existence. Similar sentiments are expressed in the Epic literature. Through institution Manu appears to have made the discovery, now known to science, that trees and plants when subjected to injury, exhibit reactions which can be photographed. Manu says: "All trees and plants are full of consciousness within themselves and are endowed with the feelings of pleasure and pain."[42]

The belief that trees have lives of their own and are associated with presiding deities required that one ask reverent permission before felling a tree, even when it was to be made into an image of deity. The wood-chopper prayed: "O tree! you are intended for the making of an image of [such-and-such a deity].I salute you. Please accept my worship offered according to injunctions... May those beings who reside here arrange for their residence else-where after accepting the oblations offered according to injunctions! Let them forgive me. Salutation to them all."[43] One can dismiss this as a relic of outdated animism, or one can see in this scenario of man and environment the *animistic expression* of a *prior and independent* value which is attributed to the tree, and which is then expressed within the spiritual conventions of the times.

Given the perception of continuity and connection between the pipal and people, tree planting was made a dharmic act. There was the promise of the expiation of sin to anyone who planted a neem tree or various fruit trees—all of which possessed nutritive and medicinal properties.[44]

The Institute of Vishnu propounds an environmentally consequentialist theory which in addition to the rewards for digging wells and giving water, includes rewards for tree planting. "He who plants trees will have those trees for his sons in a future existence. A giver of trees gladdens the gods by (offering up) their blossoms to them. (He gladdens) his guests by (giving) their fruits to them; (He gladdens) travellers with their shade; (He gladdens) the manes [ancestors] with the water (trickling down from their leaves) when it rains."[45]

If tree planting brought inherited rewards the destruction of trees incurred punishments. Kautilya levies the following fines: "For cutting off the tender sprout of fruit trees or shady trees in the parks near a city, a fine of six panas shall be imposed; for cutting off the minor branches of the same trees, twelve panas, and for cutting off the big branches, twenty four panas shall be levied. Cutting off the trunks of the same, shall be punished with the middle most amercement."[46]

Using a similar calculus, Manu declares: "According to the usefulness of the several (kinds of) trees a fine must be inflicted for injuring them; that is the settled rule."[47] The fines for injuring trees was graduated, with small fines for damaging shade trees, middling fines for damaging flower-bearing trees, and stiffest penalties for the destruction of useful fruit trees. Another factor which had to be accounted for in computing fines was the position of the tree: whether it was a boundary marker, at a crossroad, or in a monastery.[48]

The yardstick applied in these ancient systems of Kautilya and Manu is that of utility. The base line is that although all trees and plants are sacred, the human impact upon them is to be measured in terms of their usefulness. This thinking is amenable to the industrialisation of forestry, but is averse to its reduction to a commodity, which then justifies the clearance of forests to make room for more jobs.[49]

It must be admitted that the preservation and planting of trees is only a stopgap measure against global warming. But if the major peril faced by global warming is the raid pace of change which is now going on, then stopgaps can be helpful. Short of other measures, reforestation can slow the pace until other solutions can be found. But there should be no illusions. The battle is with the Goliath of Big Industry, which changes forests into fodder with the appetite of runaway consumerism. For grass-roots communities to withstand this onslaught, they must be empowered with the sort of passion that energised the Indian Chipko movement.

The Chipko movement was born as a result of the reckless destruction of the environment of the Alaknanda catchment area in the mid-Himalayas. Commercial exploitation, aided and abetted by the building of a vast network of roads for border security, involved the slashing of forests in the region. This led to soil erosion, devastating floods, and the reduction of agricultural production. The life of the villagers was perilously affected. As Chandi Prasad Bhatt describes the situation: "It was here as that the Dasholi Gram Sarajya Mandal (DGSM), Gopeshwar . . . stepped in. While doing relief work during the 1970 floods the Mandal realised that forest and land and forest and man were intricately linked. The people were also educated about it and gradually they geared themselves into the movement. Thus was born the Chipko movement."[50]

The sequel is by now well known. When the Forest Department auctioned trees as a means for acquiring foreign exchange, the brave villagers hugged the trees as a nonviolent act to stop their being cut down. The government backed off, and the local people gained their first victory. Since, then the DGSM has organised conservation camps, and has branched out beyond tree protection wider environmental concerns.

## Waste

The third problem Hindu ethics must address is that of waste. In America, it is a dilemma of our affluence; but the scourge is global. All countries face the question: "how to reduce the gargantuan waste by-products of civilisation without endangering human health or damaging the environment."[51]

The problem is multifaceted. Waste is of several types: municipal solid waste, of which the average American creates some 5 pounds a day; hazardous waste, a by-product of the chemical revolution stating in the 1930s, with approximately 650,000 commercial and industrial sources in the United States; nuclear waste, which remains toxic for thousands of years; industrial solid waste, of which 1 ton is created each week for every individual in the United States and gaseous waste, to which each person in the United States contributes 20 tons of $CO_2$ each year.[52] When the figures are added from around the world, waste becomes an urgent global dilemma. What can be done to prevent the world from wallowing in waste?

On the one hand, there are policy makers who are only guided by economic interests and who look for garbage dumps in some poor fellow's backyard. On the other hand, there are those who oppose development and nostalgically hope for the horse-and-buggy days. Hinduism's balanced scale of values (*purusharthas*) counsels neither *rape* nor *retreat.* Instead the strategy of Hindu ethics to control the volume of waste produced involves recycling, renewal, and restraint.

The waste problem has two aspects: mechanical and human. On the mechanical side, we can discuss the merits of various disposal methods, but prior to the question of disposal is the production process that creates the waste, and therefore, it is at that karmic source where we must begin. In all instances, the waste generating process must be rethought and reinvented. Forward-looking companies that have overhauled their operations have registered success. "The Minnesota Mining and Manufacturing Co. has cut waste generation in half by using fewer toxic chemicals, separting out wastes that can be refused and substituting alternative material for hazardous substances. 3M's saving last year: an astonishing $420 million.[53]

This "ascetic" approach succeeds because it approximates the natural model in which one species' waste becomes another species' means of livelihood. In nature, this cycle of *waste* and *Want* are held in balance, but in aggressive consumer societies this balance is disrupted and waste becomes *wasted* because nature is not able to sustain its reuse due to an overloaded system. By redesigning our production systems we can curb the generation of waste where it starts.

Next, with the volume of waste reduced, the remainder of waste must be recycled. Recycling is the most effective way to reduce waste. Japan has been most innovative in this area. In many cases, it has been a case of creativity born of necessity. For example, up to now, waste from Yokohama, Japan's second largest city, had been ending up in Yokohama Bay, which is filling up fast. The solution was to recycle the sewage waste. Through inventive research, the Yokohama sewage works bureau created a line of sludge products, including fertilizers, bricks, vases, paper, and business cards. Yokohama has declared its sewage system "a treasury of resources."[54]

At the consumer level, governments can prod the public to curb waste by levying a charge according to the amount of garbage they produce; to encourage recycling, households should be required to sort garbage into recyclable and nonrecyclable items.

In terms of *renewal,* there are two kinds of energy: non renewable and renewable. The first category includes natural gas, coal, oil, and peat—all being in great demand. Unfortunately, most nonrenewable sources come from fossil fuels that emit gases that contribute to smog, urban air pollution, and acid rain, and that are enormously productive of waste. It is, therefore, imperative for society to switch to renewable sources of energy by making greater use of water, geothermal heat, solar radiation, and wind.

On the principle of *restraint,* the concept of *bhu-rina,* introduced in this chapter, elevates conservationism above consumerism. We are to put back into the system what we take out, in order to maintain its ecological balance. Manu makes the point well: "As the leech, the calf, and the bee take their food little by little, even so must the king draw from his realm moderate annual taxes."[55]

Rich countries take more out of the earth than they replenish, and poor countries are playing "catch-up." The United States is the greatest consumer of them all, producing the most amount of waste. Clearly, we need to curb our voracious appetite for energy with stringent new energy efficiency regulations and widespread conservation practices. Gandhi maintained that nature's supply is always adequate when demands are managed intelligently.

## Population

The fourth environmental problem Hindu ethics must address is the problem of population. The two giants in this areas, casting colossal shadows over 40 per cent of the world's population, are China, first, and then India. But those standings could be shortly changed, since India is moving toward topping China as the most populous nation in the world by the year 2025. Earling Dessau, representing the United Nations Fund for Population Activities, released the UNFPA's annual report in New Delhi on April 29, 1992, which announced that "medium growth projection shows India overtaking China by 2025 or 2030 as the world's most populous country with a population of over 1.4 billion people."[56] An earlier projection had set China's population by 2025 at 1.5 billion, but that figure has been reduced to 1.4 billion, taking into account China's stringent population control policies.

However, being a democracy, India cannot resort to the strong-arm policies of its neighbour, as the ouster from office of the late prime minister, Indira Gandhi, attests. But democratic pride does not make the requirements for feeding, housing, clothing, and gainful employment by 2025 anything less than staggering. The UNFPA report noted that India's progress toward improving the equality of life had been slower than might have been expected. About 350 million people in India live in absolute poverty—more than its population at the time of independence in 1947.

The UN report also highlighted achievements of the southern state of Keral, in contrast to the five large states in the north where 40 percent of India's people live. "Where Kerala has matched the human resource development of the very best performers among developing countries, the five northern states have been among the worst. And the place of women is central to the difference." the report said. [57]

We hark back to a speech of the late prime minister, Rajiv Gandhi, on the eve of Republic Day, 1987, in which he said: The real benefits of development can be had in the long run only if we tackle our population growth cffetively. In this task, the active participation and cooperation of the people have to enlisted. Family welfare should be developed into a mass movement. Social and attitudinal changes have to be brought about through sustained endeavours. Voluntary organisations have to supplement the efforts of the Government in this regard.[58]

Responding to Rajiv Gandhi's challenge, the question arises: What part can Hindu ethics play toward supplementing the effort of the government, especially in bringing about necessary social and attitudinal changes?

First, the concept of *dharma* or morality is a dynamic notion in Hindu ethics, open to processes of change as different situations arise. The *dharma* of the Vedas was to beget "ten sons." That made sense in times when warfare, an agricultural economy, and high rate of infact mortality were the order of the day. For most of India's history, these exigencies have prevailed, and hence the practice of having large families has become part of a hallowed tradition. But today the same *dharmic* principle of welfare dictates a radical shift in attitude, both toward fertility and toward fertility as a requirement of *dharma.* Bluntly stated yesterday's *dharma* is today's *adharma.* "The rules of *dharma* are the mortal flesh of immortal ideas and so are mutable."[59]

Today it is ethically irresponsible to say, "God sends children into the world," and that "man should not interfere with the will of God." In Hindu ethics, that which distinguishes civilisation from the jungle is the capacity of persons to control nature, including their own. Sexual restraint is the infrastructure of the *ashrama* scheme, in which procreation is limited to the householder stage. It is also the rationale for sanctions against sexual intercourse on a number of auspicious days and seasons.

But, having acknowledged the place of restraint in family planning, it is not our purpose to isolate this age-old Hindu virtue as a method of birth control. That was Gandhi's solution to the problem. "I want limitation of population," he declared, "but the method which we adopt is the method of abstinence, austerity, and self-control." Radhakrishnan opposed Gandhi on this point: There is no doubt that it [abstinence] is the best method, but, I should like to ask, whether it be by self-control or abstinence, or austerity of living that limitation of population is brought about, is not that interference with nature? If we do not control ourselves we will produce more, and if we do control ourselves, we will produce less. There is an interference with nature even there, and so far as we are concerned most of us are human beings striving to be saintly, but have not yet become saints.[60]

The last remark of Radhakrishana reflects the thinking of another Hindu sage, Rabindra Nath Tagore. Opposing Gandhi's plea for abstinence, he said: "I believe that to wait till the moral sense of man becomes a great deal more powerful than it is now and till then to allow countless generations of children to suffer privations and untimely death for no fault of their own is a great social injutice which should not be tolerated."[61]

Indians need control, but they need contraceptives first. Seshagiri Rao correctly states: Generally Hindus show little resistance to the idea of birth control. There is no objection to contraception on religious grounds. Brihadaranyaka Upanishad spells out a method of birth control for a man "who desires a wife, but does not want her to conceive" (VI.4.10). A temperate exercise of sex instinct by the householder is recommended in the shastras...; and contraceptions help in avoiding unwanted and undesirable pregnancies. Hence, contraceptions is generally acceptable to the Hindus and is to be encouraged in the interests of domestic felicity and welfare of society.[62]

Family planning should involve more than the

distribution of contraceptives. One mistake of the past has been to rely on contraceptives in isolation. What needs to be emphsised is the integration of family planning with family welfare, environmental sanitation, child nutrition, literacy, and social and economic independence for women. This "*package*" approach versus the conventional piecemeal approach is ethically more sound because it does justice to Hinduism's concern for the whole person and for the environment in which he or she lives.

Two remaining factors that are essential to a population ethic involve marriage and offspring. Traditional Hindu marriage has a fivefold purpose: pleasure (*rati*), parenthood (*prajati*) companionship (*sakhya*), sacrificial service (*yajna*), and spiritual blis *(ananda)*.[63] Because of the historic need for large families, it is understandable how the primary meaning of marriage developed around the notion of *prajati* (parenthood). This identification of marriage with procreation has tended to define the wife's role as primarily biological. The young bride only becomes a fully fledged member of the husband's family when she produces a child, preferably a son. Traditionally, without this her traditional status has been in jeopardy. Today, with the need to limit the size of families, the social and spiritual purposes of marriage ought to be cultivated around the ideal of *sakhya* or companionship. A wife is a friend, not a factory. This sentiment is expressed in the words of the groom to the bride in the marriage hymn of the Rigveda, "I take thy hand in mine for happy fortune that thou mayst reach old age with me thy husband."[64]

Finally, in developing a population ethic, Hindus are obligated to face honestly the issue of equality of offspring. Prejudice and injutice run deep. They are hallowed by religion and flourish in an ethos that is patriarchal to the core. Success in this area could can not only a moral victory but also a reduction of the birthrate by a substantial percentage. Signs of hope can be read on billboards of family planning agencies that promote the "value of female babies in an effort to open a new perspective for many Indian couples who feel their family is incomplete unless they have atleast one male child."[65]

The need for a new perspective grows out of Hinduism's traditional preference of male offspring. The religious basis for this preference is the belief that the son rescues the souls of departed ancestors from hell. Manu says: "By a son one conquers the worlds, by a son's son one attains the infinite, by the son of a sons' son one attains the region of the sun. Since, a son succours his father from the hell called *Put*; hence, the self-begotten one (Brahma) has called a son, *Putra* (lit., deliverer from the hell of *Put*)."[66]

The status of the eldest son in the family is unique. He has the right to offer the funeral cake *(pinda)* at the time of the *shraddha* ceremony marking the anniversary of the death of the father and ancestors.

Responsibility for performance of rites for the forefathers stems from the notion that the sins of commission and omission done by parents are visited upon their children. Retribution can be averted by means of the *shraddha* rites. These household ceremonies are conducted every month for one year after a death. During this time, it is believed that the deceased father moves in the air, not having been admitted to the company of the forefathers. At the conclusion of the year, a ceremony is performed by which the deceased is admitted to the assembly of the forefathers and thereby becomes a *pitri*. Manu states that the first-born son on whom the father passes his debt(*rinam*) and by whom he gains immortality (*anantyam*) is the only child begotten for the sake of *dharma (sa eva dharmajah putrah)*: all others are the fruit of passion (*kamajah*)[67]

The importance of *shraddha* rites elevates the role and status of the son and devalues the daugher. When the dowry, which the father must give upon the marriage of the daughter, is also calculated, her liability increases as compared with a son, who is an asset to his father in every way—spiritual, social, and economic. No wonder there are always accounts of female infanticide, especially among low-caste families.

It is time that Hindus reevaluated the unethical accretions which have grown around their popular eschatology. Hindu beliefs are held on progressive levels of sophistication, but *shraddha,* though philosophically low, is not recognised as such because it caters to a host of non religious factors that assume male superiority and are therefore difficult to eject in a patriarchal society.

Philosophically viewed, the declared function of a son "as an instrument for cleansing his ancestors in the funeral rites" is purely magical. The private performance of *shraddha,* does outside the sight of eunuchs, outcastes, heretics, or pregnant women for fear that it might be rendered "unclean," is both magical and superstitious. It demeans Hindu ethics to be associated with beliefs about collective guilt; with the belief that doing good deeds will acquire merit; and with the notion that offerings (excluding "unconsecrated grain or lentils, gourds, garlics, onions, or red vegetable extracts," and salt) somehow nourish the ethereal bodies of the departed ancestors and enable them to perform works of merit that will advance them toward their goal of union with *Brahman.*[68]

Hindu ethics is more rationally served by the philosophic belief in *karma,* which asserts that each person makes his or her own heaven or hell. Moreover, our deeds are private and therefore cannot transfer for good or ill to other persons. Most importantly, a religion that views life unitively is ethically inconsistent when it elevates the salvific superiority of one sex above the other. It is strange that a religion that establishes the worship of animals and the veneration of trees should discriminate against its own daughters. In today's world of exploding populations, there is a more ethical way to affirm the continuity between generations and to honour and immortalise the names of beloved ancestors. This is achieved by performing food deeds—the most dharmic of which may be limiting the size of one's family.

## Nuclear Proliferation

Finally, we examine the Hindu ethical perspective on the most threatening of all environmental problems—the danger of nuclear war. On the one hand, it appears that with the dismemberment of the Soviety Union the fear of nuclear catastrophe has ended and there is victory for common sense. At last, we have the opportunity to dismantle the giant military-industrial machine, and turn swords into plough shares. On the other hand, as discussed earlier various dictators are scrambling to acquire nuclear power for military purposes.

In addition, the demons of nationalism and terrorism threaten international stability. Even in America, we are hearing nationalistic voices calling for the government to put America first by attending to domestic causes instead of dispensing foreign aid. This inward turning is comprehensible, considering the burden of the special role the United States has had to play throughout this century, at immense material and human cost. But America's current dilemma is that she can no more remove herself from her traditional role of global responsibility as she can put her trust in the military might that makes her the only superpower in the world today.

Some argue that it was precisely her swift and massive power in the Persian Gulf War that routed a brutal dictator in "the Mother of all battles." They want America to be a policeman to the world, using her armed forces to keep rebel nations in their place. Actually, the Persian Gulf War was a victory for the world community, by whose mandate the allied forces joined in battle. The Gulf War served as the defining event of a possible new world order in which the United Nations would assume a new and vigorous role as an instrument for peace.

The Gulf War also demonstrated that there are deep divisions between people that defy easy solutions. It showed the fundamental need for changes in consciousness between people and nations. And the doomsday sight of hundreds of uncontrollable oil fires under blackened skies, with birds and beasts sinking in seas of sludge, served as apocalyptic images of a possible Armageddon that could vaporise the planet. All of this calls for a reassessment of values. As former president Mikhail Gorbachev has revealed, it was "a reassessment of values" that "opened the road for deep transformations" in the Soviet Union. [69]

Hindu ethics is based on a set of assumptions which transcend the natural divisions between people and people, and between mankind and nature. Its *ecumenism is existential and environmental.* It sees humanity in nature and nature in humanity.

In his chapter "Vedanta in the Nuclear Age," Karan Singh argues convincingly that humankind is at the crossroads and is groping for a new philosophy to replace the old. This is taking place at a time when the supreme peril people face is not from without but from within. "There has been a tragic divergence between knowledge and wisdom, and from deep within the human psyche there has developed a terrible poison that threatens not only our now generation but countless generations yet unborn; not only out own race but all life on this planet."[70]

Singh goes on to delineate Vedantic insights which provide a framework to support the emerging global consciousness on Planet Earth. These insights are: the *Brahman* is the divine force that pervades the entire cosmos: that Brahman resides in each individual as the *Atman:* that by virtue of their shared spirituality, all humans are members of a single family; that solidarity must extend beyond human welfare to include all creation.[71]

The principles of *unity, interdependence, interconnectedness,* and *restraint* incorporated in these Vedantic assumptions might have sounded naively utopian in an earlier age. Even as recently as the forties, when Wendell Wilkie published his *One World,* the notion seemed like a dream. But as we witness nuclear weapons proliferate before our eyes, yesterday's dream becomes today's nightmare. Today's shrunken world cannot permit the luxury of national particularism. The entrenched status quo that has lasted interminably is now being pressured by the forces of modernity. Their message is clear: *interdepend or perish!* Our technological civilisation leaves us in no doubt: interdependence is not a theory, but a fact; not an ideology, but a condition. Technologically, economically, politically, and morally, interdependence in inter-national relations is inescapable. No more can any single nation be the master of its own fate and the captain of its own soul.

A dramatic example of interdependence is the current AIDS epidemic. The global spread of this disease confounds national boundaries. It demands global remedies. No single nation or group of nations, no matter how technologically advanced, possesses the capabilities or resources to arrest the deadly virus independently. What was first dubbed as an "African problem," then a "Haitian problem," then a "gay problem," has now surfaced, like a hydra-headed monster, as a global problem.

AIDS is the latest in a long list of other problems—poverty, hunger, environmental decay, resource depletion, population pressure, international terrorism, and nuclear proliferation—that once were neatly and logically discussed within the closed categories of the "life-boat" ethic. But in an interdepenent world there are no lifeboats. We either sink or swim—together. Like the king who discovered that his royal command could not stop the tide, no country can place "No Entry" signs upon its national border against any of these problems—most especially the threat of nuclear suicide and terricide.

The general idea behind Hinduism's Environmental ethics is that the individual *atman* is one with the universal *Brahman.* This *Brahman* force is manifest uniformly in the divinities of heaven and in human, animal, and plant life on earth. All of these entities live an apparently independent existence, but they all emanate from *Brahman* and are finally reabsorbed into it. *Brahman* itself is infinite, and is therefore greater than the sum of all its manifestations, past, present, and future.

This belief in *Brahman* provides the philosophic basis for the Hindu's veneration of the natural world. The universe appears to be material, but actually it is the universal consciousness or *Brahman.* Since all is one, the conquest of nature cannot be true to reality, and our sense of separations is the product of ignorance.

This calls for self-knowledge. Humans cannot act ethically toward nature as long as they are ignorant of themselves. Lacking our own sense of universal identity, we cannot identify with the trees and the mountains, nor can we feel empathy for the beasts of the fields. Nature is perceived as empty because we are empty. We manipulate nature because we manipulate ourselves. Our loss of our relation to nature is a corollary of the loss of our own selves. William Wordsworth recognised this correlation between our inner world and the outer world, and expressed it in his sonnet:

> The world is too much with us; late and soon,
> Getting and spending, we lay waste our powers;
> Little we see in Nature that is ours:
> We have given our hearts away, a sordid boon!
> This sea that bares her bosom to the moon,
> The winds that will be howling at all hours,
> And are up gathered now like sleeping flowers.
> For this for everything, we are out of tune.[71]

The rallying message of Hindu ethics is that harmony is already here: That we do not have to create it—only discover it. Since, *Brahman* and Nature are one, we must see the Supreme Being in the whole world, and the whole world in Him.

## REFERENCES


1. Norman Cousins, "Cleaning Humanity's Nest" *Saturday Review*, March 7, 1970.
2. *Ibid.*
3. *Ibid.*
4. Aldo Leopold, *Readings in Conservation Ecology*, ed. George Cox (New York: Appleton-Century-Crofts, 1969), p. 585.
5. Ainslie T. Embree (ed.) *The Hindu Tradition*, (New York: The Modern Library, 1966), p. 9.
6. *Ibid.*, p. 16.
7. *Rigveda* 146.5 Ralph Griffiths, trans., *The Hymns of the Rigveda*, ed. J.L. Shastri, New Delhi, Motilal Banarsidass, 1973, p. 641.
8. See R.S. Bhatnagar, "The Emergence of Basic Ethical Concepts: First Book of Rig Veda," unpublished paper.
9. *Atharvaveda* 12.14

10. Atharvaveda 12. 35.

11. *Ibid.*, 12.12.

12. S. Dasgupta, *A History of Indian Philosophy* (Cambridge: Cambridge University Press, 1969), Vol. 1, p. 42.

13. Brh. Up. 4.4.22, in R.E. Hume, *The Thirteen Principal Upanishads,* 2d ed., rev. (London: Oxford University Press, 1971), p. 143.

14. See Tait. Up. 2.8; Maitri Up. 6.17; Brh. Up. 5.15; Isha Up. 16.

15. Isha Up. 1.1., in Hume, *op. cit.*, p. 362.

16. SK. 21.

17. SK. 57.

18. M. Hiriyanna, *The Essentials of Hindu Philosophy* (London: Allen and Unwin, 1967), p. 119.

19. Brh. Up. 2.5.15, in Hume, *op. cit.*, p. 104.

20. Bhagavadgita, 2.28.

21. Fritjof Capra and David Steindl-Rast, *Belonging to the Universe* (New York Harper, 1991), p. 83.

22. Eliot Deutsch, "Religion and Ecology: Ancient India and Modern America," Paper presented at conference on Religion and Ecology, Cromwell Crawford, Director, University of Hawaii, Honoloulu, 1968.

23. Brh. Up. 4.4.5, in Hume, *op.cit.*, p. 140.

24. Deutsch, *op. cit.*

25. Gore, *op.cit.*, p. 4.

26. Manu 6.35, 37, also Yaj. 3.56; Bau. 2.34; Apa. 2.24.

27. Walter H. Maurer, *Pinnacles of India's Past* (Philadelphia: John Benjamins Publishing Company, 1986), p. 219.

28. In Margaret and James Stutley, *Harper's Dictionary of Hinduism,* ed. Margaret and James Stutley (New York: Harper and Row, 1977), p. 234.

29. Maurer, *loc. cit.*

30. Ian Barbour, "Environment and Man," in *Encyclopaedia*

*of Bioethics*, Vol. 1, *op. cit.*, p. 370.

31. Gore, *op. cit.*, p. 50.
32. Michael D. Lemonick, "Feeling the Heat," *Time*, January 2, 1989, p. 36.
33. Leopold, *op. cit.*
34. Linden, *op. cit.*, p. 34.
35. Advertiser News Sevice, "Washington Taking the Heat at Earth Summit," *Sunday Star-Bulletin and Advertiser*, June 1992, p. A.25.
36. Linden, *op. cit.*, p. 34.
37. Charles, P. Alexander, "On the Defensive," *Time*, June 15, 1992, p. 35.
38. Karan Singh, *Hinduism*, 2nd rev. ed. (Delhi: Ratna Sagar, 1990), p. 118.
39. *Young India*, June 18, 1921.
40. Advertiser News Service, loc. cit.
41. Philip Elmer-Dewitt, "Rich vs. Poor," *Times*, June, 1, 1992, p. 43.
42. Manu 1.49.
43. Brihatsamhita 59.
44. Varaha Purana, 172. 39.
45. Vishnu, 91. 4-8.
46. 111.19.197.
47. Manu 8.285.
48. Ibid., note on vs. 285.
49. Michael Weisskopf, "Quayle, Gore Now Point Men for Environment," *The Sunday Star Bulletin and Advertiser*, p. A30.
50. Chandi Prasad Bhatt, "The Hug That Saves," in Gadgil, *op. cit.*, p. 17.
51. Langone, *op. cit.*, p. 44.
52. See Gore, *op. cit.*, p. 146.

53. Langone, *op. cit.*, p. 45.

54. "Japan Gets Creative in Sewage Business," *Chicago Tribune*, June 7, 1992, Section 7-7B.

55.. Manu 7.129.

56. Reuters, "U.N.: India Will Top China as Most Populous Nation in World by 2025," *Honolulu Star-Bulletin*, April 30, 1992.

57. *Ibid.*

58. "President's Message," *Indiagram*, January 26, 1987, p. 2.

59. S. Radhakrishnan, Quoted in S.Chandrashekhar, "How India Is Tackling Her Population Problem," *Foreign Affairs*, October 1968, no. 1, p. 145.

60. S. Radhakrishnan, *Planned Parenthood* (New Delhi: Central Planning Institute, 1969), p. 11.

61. H. C. Ganguli, "Religion and the Population Explosion," *Religious Pluralism and the World Community*, ed. E. J. Jurji (Leiden: E. J. Brill, 1969), p. 208.

62. Seshagiri Rao, "Population Ethics: A Hindu Perspective," *Encyclopaedia of Bioethics*, Vol. 3, *op. cit.*, p. 1271.

63. Manu 9.137, 138 in *Dharam Shastra*, Vol. 5, ed. M.N. Dutta (New Delhi: Cosmo Publication, 1979), p. 334.

64. Rigveda 10.85.36.

65. Burdett, "Ambitious Targets Highlights India's Population Strategy," *Popline*, December 1986, p. 4.

66. Manu 9.137, 138.

67. *Ibid.*, 9.107.

68. Stutley, *op. cit.*, p. 284.

69. Mikhail Gorbachev, "View from Washington," *Honolulu Star-Bulletin*, April 30, 1992.

70. Karan Singh, *op. cit.*, p. 44.

71. *Ibid.*, pp. 45-47.

72. William Wordsworth, *The Golden Treasury*, ed. F. T. Palgrave (Nelson and Sons, n.d.), p. 310.

# 11

# ENVIRONMENTAL ECONOMY AND JUSTICE

Environmental Economics has so far enjoyed some successes and has encountered some problems in developing methods to measure the costs and benefits of alternative environmental outcomes. The article begins by explaining that environmental economics is a *normative* as distinct from a *positive* science insofar as it proposes goals (principally *social welfare*) for regulatory policy. The article then introduces the concept of *preference,* which economists use as the building block of social welfare. The ideal of a perfectly competitive market then introduces the thesis the *market failure* causes a divergence between the *private costs* of a transaction (i.e., costs reflected in prices paid by parties to that transaction) and its *social costs* (i.e., those passed on to those not party to the exchange). There follows a brief introduction to the kinds of market failure, such as *externalities, public goods,* and *transaction costs,* thought to justify governmental interventions in or regulation of markets. The article then reviews techniques environmental economists have proposed to help public officials to regulate environmental resources to correct these and other forms of market failure.

The article also identifies problems environmental

economics still must resolve. The first of these concerns the extent to which the concept of *welfare or well-being,* which economists define in terms of *willingness to pay*, can be related to goods or goals, such as human happiness or contentment, as these concepts are understood by noneconomists. (Critics contend that environmental economists use "welfare" in a technical sense that bears no relation, conceptual or empirical, to what we ordinarily understand by that term.) A second difficulty relates to the possible circularity that arises in the description of preferences. Because preferences are unobservable they are inferred from descriptions of behaviour. To describe behaviour, one must already make assumptions about the preferences that underlie it. Third, critics question whether economists can assemble information more accurately than markets, however flawed; here the problem of disagreement among experts ("dueling cost-benefit analyses") arises. Finally, in empirical studies, people report that they value environmental goods for ethical, cultural , or spiritual reasons unrelated to their own well-being. It is not clear how these "commitment" values, if unrelated to the individuals' welfare, can enter into the social welfare calculus on which environmental economists believe public policy is to be based.

In 1981, President Reagan ordered all federal agencies to refrain from undertaking major regulatory actions "unless the benefits to society from the regulation outweigh the costs." Executive Order 12291, published in the Federal Register in February 10 of that year, sought insofar as possible to replace the cumbersome regulatory cotraptions created by statute with scientifically based techniques of policy analysis environmental economists had developed. Environmental economics is the branch of welfare economics that is concerned with helping society to use natural resources in ways that provide the greatest net benefits to society as a whole. Many environmental economists saw Executive Order 12291 as a mandate for the objective evaluation of environmental assets a guide to public policy.

"The basic premises of welfare economics," as Stokey and Zeckhauser observe, "are that the purpose of economic activity is to increase the well-being of the individuals that make up the society, and that each individual is the best judge of how well off he or she is in a given situation." Because each individual is the best judge of his or her own welfare gains or losses, the individual's preferences, measured by his or her *willingness to pay* to satisfy then, indicates the welfare change—gains or losses—the individual expects as a result of a particular environmental decision. Economist A. Myrick Freeman defines "the benefit of an environmental improvement as the sum of the monetary values assigned to these effects by all individuals directly or indirectly affected by that action." Environmental economists Joseph Seneca and Michael Taussig concur, "The *benefit* of any good or service is simply its value to a consumer."

According to the science of welfare economics, as economists Edith Stokey and Richard Zeckhauser explain, "the purpose of public decisions is to promote the welfare of society." They add that "the welfare levels of the individual members of society are the building blocks for the welfare of society." From this perspective, a regulation is rational—it promotes the welfare of society—only it confers on members of society in the aggregate benefits in excess of costs. Because the benefits and costs may well accrue to different individuals, welfare economists recognise two fundamental values in terms of which regulatory policy may be justified. The first is economic efficiency, which is to say, the extent to which the benefits of a policy exceed the costs for society as a whole. The second goal is *equity,* that is, the extent to which the distribution of costs and benefits is equitable or fair.

Seneca and Taussig define efficiency as the "maximum consumption of goods and services given the available amount of resources." They characterise equity as "a just distribution of total goods and services among all consumer units." As economist Arthur Okun has written, "To the economist, efficiency implies getting the most out of a given input." He explains: "This concept of efficiency implies that more is better, insofar as the 'more' consists in items people want to buy." Roughly speaking, efficiency in the allocation of resources makes the social pie as big as possible; equity in the distribution of resources fairly divides that pie.

Many economists, such as Okun, propose that economics as a science properly addresses issues of efficiency, while equity concerns may be considered by others, such as politicians and philosophers. Other economists, including Burton Weisbrod, have argued that because preferences over the distribution of wealth seem indistinguishable from other preferences, they might be assimilated into the analysis of overall welfare. Economists could derive social policy scientifically from the preferences of individuals, whether these preferences involve the allocation of resources or the distribution of income. On the other hand, equity concerns may be said to differ from other preferences because they have a public, political, or objective dimension. They reflect what an individual believes is good from the perspective of society as a whole—what is objectively right—not desires about his or her private consumption opportunities. This would seem to be true, however, of many other preferences concerning society, politics, or the environment. Later we shall consider the problem economists encounter in dealing with moral or "commitment" values that express the individual's view of the "good society" rather than of the "good life."

## Positive and Normative Science

Environmental Economics, as an application of the more general theory of welfare economics, presents itself as a normative discipline. As economists explain, its goal has been "to evalutate the social desirability of alternative allocations of resources" and "to formulate propositions by which we may rank, on the scale of better or worse, alternative economic situations open to society." As a normative discipline, environmental economics distinguishes itself from the "positive" study of the functioning of an economy. "Positive economics is the study of what is," a textbook explains, "normative economics is the study of what ought to be." To describe the way an economy works, offering predictions without evaluations, is the province of positive economics. "However, a study to determine the kind of regulation we ought to adopt for a particular environmental problem is a case of normative economics because it...involves value judgements."

Normative welfare and therefore environmental economics bases itself on "one fundamental ethical postulate"—that the preferences of individuals are to determine the allocation of resources."In this framework," Economist Alan Randall notes, "preferences are treated as data of the most fundamental kind. Value, in the economic sense, is ultimately derived from individual preferences..." Environmental economist A. Myrick Freeman III explains the relation between individual preferences and social value: Since, the benefits and costs are valued in terms of their effects on individuals' well-being, the terms "economic value" and "welfare change" can be used interchangeably. Society should make changes...only if the results are worth more in terms of individuals' welfare than what is given up by diverting resources and inputs from other uses.

David Pearce, a prominent British environmental economist, argues that the only values that should count in Environmental policy are subjective consumer preferences. He writes: "Economic values are about what people want. Something has economic value—is a benefit—if it satisfies individual preferences." Three economists observe, "Economic value for the society arises from individual preferences which are taken to be exogenously given." Michael Bowes and John Krutilla similarly argue that the goal of environmental policy is to allocate environmental resources "to provide the greatest discounted net present value from the resulting flow of goods and services." To say that environmental economics is a normative science is to note that it prescribes the goal of environmental policy, namely, to give the members of society as much as possible of the things that they believe will improve their welfare, which is to say, the things they want to buy.

## Public Policy

It is important to recognise that the framework of welfare economics is not the only one that has been proposed for justifying environmental policy. James G. March, in his *Primer on Decision-Making* (1994), distinguishes between two conceptions of rational collective choice. He distinguishes between utilitarian or economic (presence-based) and deontological or Kantian (principle-based) decision-making. The *utilitarian* approach, which underlie the science of environmental economics, chooses among given alternatives "by evaluating their consequences in terms of prior preferences." Environmental economists appeal to utilitarian moral theory to justify policies intended to allocate resources to those willing to pay the most for them—and who thus may be supposed to benefit the most (or be able to benefit others) by possessing and using those resources.

James March contrasts utilitarianism with a rule-based or *deontological* approach to policy making. When decision-makers adopt the deontological perspective, according to March, they "pursue a logic of appropriateness, fulfilling identities or roles by recognising situations and following rules that match appropriate behaviour to the situations they encounter." In this context, individuals typically do not ask, "what outcome will most benefit me as an individual?" but, "What do we believe is appropriate for society as a whole, given our shared principles, beliefs, and commitments?" From this perspective, political processes and institutions are supposed to provide the context in which citizens try to convince each other of their views and exact them by majority vote within limits set by political, personal, and civil rights. "The reasoning process," March writes, "is one of establishing identities and matching rules to recognised situations."

What distinguishes these two approaches most fundamentally is the way they regard or interpret disagreement among members of society. The utilitarian understands disagreement as conflict or competition for the use of scarce resources. On this view, resources must be allocated to satisfy "given" wants and desires. The ways to resolve conflict, then, is to allocate resources in ways that provides individuals the most of the goods and services they wish to buy.

The opposing approach, which is associated with the eighteenth-century German philosopher Immanuel Kant, analyses disagreement in terms of the logical opposition of moral or political beliefs. In answering questions such as, "what sort of people are we?" and "What do we stand for as a nation?" Individuals express opposing views of the common good rather than reveal their preferences with respect to their personal good. Some citizens may assert that society should maximise welfare, but others my hold, for example, that we are committed as a nation to value nature for its own sake or as a heritage for future generations. From the Kantian perspective, these views about the public interest are not to be assimilated to the self-regarding welfare concerns of individuals. They are to be understood as political assertions about the appropriate goals of public policy.

The two frameworks for decision-making differ, then, in the way they conceive the decision-maker. In the utilitarian perspective, the decision-maker is the individual whose judgements about his or her welfare with respect to consumption opportunities result in sets of preferences over environmental goods, services, and outcomes. The individual functions as the location at which subjective wants or desires are found. The problem of public analysis is therefore to ascertain the preferences of the individuals in the client society and determine the most efficient way to allocate resources to satisfy them. Many economists believe that all rational individuals should agree of this approach, and therefore their hypothetical consent to it may be assumed, because it promises to give them the most of the goods and services for which they are willing to pay.

The opposing deontological approach may accept the precepts of welfare economics as far as the subjective consumption desires of individuals are concerned. The deontologist, however, emphasises the importance of what people believe or stand for, not just what they desire to consume. In the spirit of the eighteenth-century philosopher J.J. Rouseau, deontological theory regards the individual as a source of judgements, views, or arguments about public policy, that is, as defending conceptions of the common good. In other words, the individual is conceived as a person whose views about public policy are to be weighed on the merits, not as a container of preferences that are to be priced at the margin. Of course, if everyone agreed that the goal of environmental policy is to maximise "welfare," as economists understand that term, there would be little reason for political as distinct from economic decision-making. Because citizens differ in their views and conceptions of public policy, however, they must rely on democratic institutions and processes to sort out disagreements. Democracy is the appropriate political system for those who agree to disagree about the normative and conceptual foundations of public policy.

The deontological views entertains the possibility that citizens will espouse views of the good society and not just preferences concerning their own good. The utilitarian approach, in contrast, regards all such values as subjective preferences directed to personal will-being. Within theoretical framework of welfare economics, only views about policy that are scientifically valid, namely, those of welfare economists, are to be accepted on the merits. This does not mean that environmental economists dismiss the views of others, for example, people who believe that a good society would not trade off the existence of species for economic gain. Rather, these views about public policy are assumed by economists to express a special kind of want or subjective preference—say an "existence" value—that must be factored into the cost-benefit analysis on which policy is to be based.

Because environmental regulation in our democratic society has not been poetry in motion—regulations can impose great economic costs for modest benefits—environmental economists have welcomed Executive Order 12291 and other administrative attempts to establish the goals of public policy on rational scientific grounds. As one textbook suggests, the move from political to scientific judgement is needed because politicians seek their own good, while economists are dedicated to the good of society as a whole, "It is the politician's job to compromise or seek advantage" while economists and other scientists "produce studies that are...as objective as possible."

## Environmental Economics and Market Failure

There is a great deal to be said in favour of free markets as instruments for allocating environmental resources. First, we know from experience that free market economies have done a far better job protecting the environment as well as producing wealth than socialist economies, such as the former Soviet Union. Prosperity—for example, full employment, low inflation, low interest rate and so on—is good for the environment insofar as it produces wealth needed to control pollution and preserve nature. Few environmentalists, indeed, would endorse centralised planning as an alternative to a free market economy.

Second, free markets encourage individuals to take responsibility for their own choices rather than to seek favours from or lay blame on an all-powerful government. Third and most important, markets function marvelously well in gathering information not available to any public official or agency. The political economist Friedrich Hayek argued that the principal advantage of a market is not the organisation of self-interest but the organisation of knowledge. Markets thus address "the constitutional limits of man's knowledge and interests, the fact that he *cannot* know more than a tiny part of the whole society."

Political economists in the tradition of Hayek doubt that centralised planners or scientific managers can either master enough information or detach themselves so thoroughly from their own interests and prejudices to allocate resources better than individuals will themselves thorough voluntary systems of exchange. Hayek notes that the practice of experts to refer to the wants or preferences of individuals either as "data" or as "given" (or even by the pleonasm of "given data") other leads economists to assume that this knowledge exists not merely in dispersed form but that the whole of it might be available to some single mind. This conceals the character of competition a discovery procedure.

Environmental economists, as we have seen, concern themselves primarily with the goal of allocating resources to those who are willing to pay the most for them in order to maximise social welfare. It is important to understand that this goal—*allocative efficiency*— bears little or no relation to the reasons, such as those just listed, that lead many environmentalists to favour a free market system. There is no connection, empirical or conceptual, for example, between allocative (or micro-economic) efficiency and social prosperity. Prosperity, that is, the macroeconomic performance of a society—expressed in measurements of employment, inflation, interests rates, balance of trade, and so on—depends on such things as fiscal and monetary policy, the education and skill levels of citizens, their technological progress and prowess, and so on, far more than on the efficiency of resource utilisation for consumption. Some economists believe (flowing Marx) that efficient markets must drive down prices to costs, eliminating profit and therefore leading to massive unemployment. Others contend, on the contrary, that involuntary unemployment is impossible within efficient markets, because wages will fall sufficiently to clear the labour pool. In fact, nobody knows how to make sense of the macroeconomic effects of allocative efficiency, if there even are such effects. Accordingly, environmental economists, as a rule, make no claims about the relation between macroeconomic efficiency, which is their concern, and macroeconomic performance.

Environmental economists, again differing from Hayek and other political economists, put little store by the ability of free markets to gather and utilise information. On the contrary, environmental economics point out that only under certain ideal conditions that rarely if ever hold will decentralised market exchange allocate resources efficiently. Under these ideal conditions, the *private costs* of production and consumption equal the *social costs.* This means that the prices both firms and individuals pay reflect the full costs (including the pollution costs) of producing goods and the full costs (including disposal costs) to society of consuming them. When markets meet these conditions, they are perfectly competitive and will provide exactly the mix of goods and services people most want given the limits imposed by the scarcity of resources.

The conditions that define perfectly competitive or efficient markets, however, may never fully be realised. For markets to function efficiently, property rights in everything anyone wants must be fully defined and enforced. Every would-be party to an exchange must understand its terms and conditions and the qualities of the goods and services for sale. All the costs and benefits associated with a transaction must be captured in the prices paid, so that no uncompensated costs are borne by third parties. Finally, the costs of making transactions can not be too onerous. In these ideal circumstances, everyone affected by a transaction may be supposed to benefit from it, for otherwise he or she would not have entered into it.

As environmental economists Alan Kneese and Blair Bower wrote in 1972, economic theory "developed on the presumption that virtually everything of value is suitable for private ownership with little or no 'spillover' to other persons, households, and firms when the private property is put to use by its owner." Kneese and Bower added: "Of course, it was realised that sometimes adjustments had to be made for 'market failure,' but these were implicitly, if not explicitly, regarded as minor with respect to the overall allocation." Kneese and Bower, who pioneered the application of welfare economics to environmental policy, wrote that by the 1960s, economists began to recognise that market failures were pervasive and ubiquitous. It had become clear "that the pure private property concept applies satisfactorily to a progressively narrowing range of natural resources and economic activities." Kneese and Bower concluded, "Private property and market exchange have little applicability to their allocation, development, and conservation."

As environmental economists identified more and more ways markets fail with respect to the environment, it became clear to them that the government would have to direct the allocation of environmental resources. Environmental economists therefore developed techniques, some of which we shall discuss presently, to gather the given data concerning individual willingness to pay for environmental goods and services. William Baumol and Wallae Oates, in their popular textbook on environmental policy, for example, argue that with many environmental sets such as clean air, "the process of direct negotiating and agreement will generally be unmanageable." As a result, the government must determine the prices at which these environmental assets may change hands.

In the regulatory agencies, public officials were quick to argue that whatever rules they proposed were needed to "correct" market failures During the 1960s and 1970s, in academic departments and think tanks, environmentalists joined environmental economists in playing the role of sorcerer's apprentice in the discovery of "external" or third-party "spillover" costs of actions that damaged the environment. Externalities, defined as costs and benefits markets fail to price, flooded the intellectual landscape; they appeared to overwhelm the costs and benefits markets do price, that is, those that are "internalised" or reflected in market prices. And every "spillover' or "unpriced" costs of an action suggested a reason to reallocate resources efficiently, that is, away from actual (as opposed to ideal or hypothetical) market transactions.

During the 1960s and 1970s, environmentalists found the externality argument gave them instant legitimacy within the established legal and scientific culture; therefore, they either played down their deontological—that is, ethical and cultural—values, ideals, and commitments or recast them in the language of spillovers, welfare, and willingness-to-pay. Those environmentalists who were motivated by spiritual or moral commitments opposed to a utilitarian ethic found themselves locked in the language of contemporary utility theory. Critics of the "market failure" approach charged, then, that it came primarily "from economists desperately eager to play a more significant role in environmental policy and environmental groups seeking to gain the support of conservatives."

Because market failure stories proved so easy to tell, libertarian economists in the tradition of Hayek have worried that they provide an automatic basis for public officials to second-guess market outcomes. In a book defending the ability of free markets to protect the environment, libertarian economists Terry Anderson and Donald Leal portray the cost-benefit approach of environmental economics as a pathway to collectivism. They write: To counter market failures, centralised planning is seen as a way of aggregating information about social costs and social benefits in order to maximise the value of natural resources. Decisions based on this aggregated information are to be made by disinterested resource managers whose goal is to maximise social welfare.

According to Anderson and Leal, environmental economics" stresses the potential for market failure in the natural resource and environmental arena on the grounds that externalities are pervasive." They add: "Economic analysis in general and natural resource economics in particular, have approached resource policy as if there were a 'socially efficient' allocation of resources that will be reached when scientific managers understand the relevant trade-offs and act to achieve an efficient solution." Anderson and Leal argue in the spirit of Hayek that resource managers cannot acquire "the 'correct' information about resource values" in order to find "the socially optimal allocation." Mangers often disagree, moreover, especially when hired to defend opposing positions. Opposing sides to a policy are able to present opposing cost-benefit analyses. Anderson and Leal deny, therefore, that cost benefit analysis, which they equate with "scientific management" and "centralised planning," will succeed any better than actual markets, however flawed, in allocating resources.

## The Many Faces of Market Failure

"The assertion of market failure," one economist writes, "is probably the most important argument for government interventions." Another agrees: "It is the failure of the market system to allocate and price resource and environmental services correctly that creates the need for economic measures of values to guide policy making." A third states that any governmental intervention to achieve environmental goals, should be based on the existence of a market failure (an "externality"). An "externality" is a cost that is borne by a person other than the person who caused the cost to arise...Absent market failure, environmental degradation is not by itself sufficient to justify governmental action.

Among the kinds of market failure, three are the most prominent. They are *externalities, public goods,* and *transaction costs.* An externality arises, for example, when a cement company casts its dust on its neighbours without obtaining their consent. It will be able to charge less for cement that a company that purchases pollution control equipment or compensates its neighbours. As a result of this uncompensated cost more cement will be produced relative to clean air than people want. Anderson and Leal summarise: market failure is said to result when any benefits are not captured or costs are not borne by decision makers. The existence of these externalities or third-party effects means that either too little of a good is produced in the case of uncaptured benefits or too much in the case of unborne costs. . . . Such under-or over-production is often taken as a sufficient condition for taking political control of resource allocation.

A second form of market failure arises when assets are not privately owned but belong as property to no one or to the public in common. Everyone can benefit from these *public goods* (national defense is an example) whether or not he or she pays for them. Those who own private property—a painting, for example—can charge people to see or use it and can exclude people from having access to it. The consumption of a private good involves rivalry or competition, moreover, because if one person puts the painting in his or her home, another cannot. Public goods, such as highways or sidewalks, if open to one, are open to all. Thus, individuals have an incentive to *free ride,* that is, to use the public resource while letting others pay for it. Once again, one is tempted to reap the benefits oneself while imposing on others the costs of providing or maintaining the resource.

If everyone can access a public environmental good—wild fish stocks, for example, or clean air—people tend to use as much of them as they can while paying nothing for the privilege. As a result, the resource will be overused, for example, as more and more people try to capture it for themselves. The air will get dirtier, the wild stocks of fish become depleted, roads congested, and so on, as the logic of this well-understood "tragedy of the commons" works itself out. To avoid depletion and overpollution, economists recommend that the "commons" be privatised or that a regime be established to imposed on those who use the resource a charge reflecting its value to others and to society in general.

Finally, *transaction or bargaining costs* (sometimes described in terms of "isolation paradoxes") constitute the most general reason market prices may fail to allocate resources efficiently. Suppose a cement company moves into a neighbourhood destroying its amenity by creating fumes, dust, noise, and so on. Each of the neighbours may be willing to pay an amount that, in the aggregate, would be sufficient to bribe the plant owners to locate elsewhere, but the costs of organising themselves, overcoming free-rider problems, and bargaining with the company may defeat these disorganised or isolated individuals. In general, millions of people may each be willing to pay a small amount that, when added together, would purchase many environmental goods, but no one will incur the costs of contacting everyone, taking up a collection, and bargaining for the good in question.

Many economists believe on theoretical grounds that all forms of market failure can be analysed in terms of transaction costs. Polluters would pollute only to optimum levels, for example, if their victims could costlessly organise themselves to pay them no to pollute or to demand compensation, depending on the way legal or property rights are initially distributed. Nobel laureate Ronald Coase explained that in the absence of transaction costs people will always negotiate to the same outcome (an agreement that maximises wealth) no matter how property rights are initially distributed, or even if no such rights exist. George J. Stigler dubbed this result the Coase Theorem, which he formulated as follows: " . . . under perfect competition, private and social costs will be equal."

In the light of the Coase Theorem, the principal question of environmental economics may be put this way: How would individuals allocate environmental resources in the absence for transaction costs? This is a way of asking how a perfectly competitive market would allocate those resource. There strategies suggest themselves for answering this question. First, we might try to reform institutional arrangements to minimise transaction costs, for example, by building up the infrastructure of associations and organisations in civil society capable of negotiating solutions to environmental conflicts. The field of institutional economics takes up this kind of concern. Second, environmental economists are developing complex scientific techniques, such as contingent valuation methodology, which we shall discuss presently, to determine what people would pay in a market if they could organise themselves costlessly to bid for environmental assets, goods, and services. Finally, we might accept the status quo, where people do the best they can to negotiate in markets and may also regulate behaviour through the political process.

Because there is no ideal market against which we can test predictions about what individuals would do in the absence of transaction costs, disagreements over the "correct" price of any resource are in principle unresolvable. Economists themselves incur costs, moreover, when they undertake cost-benefit analyses and employ other scientific techniques to determine how much individuals will pay for public goods in hypothetical markets. These costs mount as more and more economists are hired as expert witnesses in litigation over opposing cost-benefit analysis. For example, public officials estimate that they have spent more than $100 million to assess the natural resource damage cause by the *Exxon Valdez* incident, although the costs of determining the value even of a single oil-soaked otter may be unlimited, as long as "deep pockets" exist to support different scientific estimates. Accordingly, a lively debate has arisen whether environmental economists are better positioned than individuals themselves to overcome transaction costs and thus to determine the "true" or "correct" prices of resources.

Economist. C.J. Dahlman argues that the fact that markets fail hardly shows "that there is good reason for auming that somebody else, outside the market, can do it better." Charles Perrings, who is associated with the emerging discipline of ecological economics, observes that the cost-benefit approach rests "on an ideological assumption that the state has privileged knowledge of the...optimal outcome, and is better equipped, informationally, to achieve it." He asks whether official experts can deal with transaction costs better than individuals. Perrings adds, "The implication is that if neither the state nor any other nonmarket agency can do it better, then there is no difference between private and social cost and no cause for believing an external effect to exist."

## AIR POLLUTION

Pollution control provides a clear example of the opposition between utiliarian and deontological or rule based conceptions of collective choice. A utilitarian approach regards pollution primarily as an "external" or "social" cost of production and therefore may seek to "internalise" this cost in prices paid for goods and services that pollute. A Kantian perspective, in contrast, regards pollution as a form of coercion, invasion, or trespass to be regulated as a violation of the rights of person and property. From this perspective, pollution constitutes a tort or nuisance—like a punch in the nose. Thus, pollution control protects the rights of individuals against trespass, which is not the same thing as—indeed, it may conflict with—satisfying their consumer desires or preferences.

If society adopted a utilitarian approach, it would bring pollution up or down to "optimal" levels, that is, levels at which the benefits of controlling pollution further do not equal the costs. A Kantian framework, in contrast, minimizes pollution until the cost of further reductions become prohibitive. Both common law and public law in the United States take this deontological approach to pollution. For this reason, as economists Maureen Cropper and Wallace Oates observe, "the cornerstones of environmental policy in the United States," such as the Clean Air and Clean Water Acts, *"explicitly* prohibited the weighing of benefits against costs in the setting of environmental standards."

The Clean Air Act of 1970 required the then new Environmental Protection Agency (EPA) to set standards for a list of "criteria" pollutants to assure that air pollution would not endanger public health or welfare or harm even particularly sensitive individuals. In calling for "an adequate margin of safety," Congree pretended that safe thresholds for air pollutants could be found and that NAAQS could come under these levels. Congress knew at the time that no such levels existed; nevertheless, it set EPA to the task of ratcheting pollution down to safe standards even at considerable cost to society.

To understand this, one might consider the case of *Boomer vs. Atlantic Cement Company,* which played in the New York State courts as Congress worked on the Clean Air Act. There, plaintiff Boomer, a farmer, sought to enjoin Atlantic, a major manufacturer, from spewing dust and fumes on his land. New York State is a jurisdiction that usually grants injunctions in nuisance cases. In this instance, however, the trial court awarded Boomer compensatory damages instead, because in order to cease polluting, Atlantic, an immense plant supporting the local economy, would have to cease operations. Thus, the court effectively gave the polluter the power of eminent domain—the power to coerce the plaintiff off his land at a price set not by the plaintiff but by the court.

What the court did was efficient from an economic point of view: it allocated resources according to market-based prices. This approach establishes economically "optimal" levels of pollution—levels at which those who benefit from polluting activities could still benefit even after compensating their victims for whatever damage they suffer. These "optimal" levels, however, can be understood only in terms of a hypothetical market where people will always sell property rights to the highest bidder—not in terms of an actual market where people like Boomer just want to be left alone. By denying Boomer injunctive relief, the court departed from the common law protection of property rights to give the polluter effectively the right of eminent domain to coerce its neighbours off their land at a price set by appraisers and not by the neighbours themselves.

Many commentators believe that in enacting the Clean Air Act, Congress in part reacted against the outcome of *Boomer* to insist that the nation regard pollution not as an external cost of production but as an enjoinable nuisance or tort. The Clean Air Act, in setting air quality standard at "no-rick" levels, treated persons as ends-in-themselves rather than as means in the overall maximisation of social wealth. Executive Order 12291 and the cost-benefit criterion it imposed then sought to replace the health and safety based imperatives set by statute with the efficiency goals suggested by welfare economics.

## THE VALUE OF BIODIVERSITY

Congress passed the Endangered Species Act (ESA) in 1972 to prevent or at least greatly to slow the extinction of species in the United States. The Fish and Wildlife Service has since listed about 956 species as endangered in the United States. No market player—for example, food, pharmaceutical, or seed company—has acted to protect habitat in the expectation that one or another of these creatures might prove useful. Are there economic benefits that justify the costs of species preservation?

Everyone knows that biodiversity as a whole or in general is essential to all life (indeed, it is all life), but this is irrelevant to economic value, which is always computed at the margin, that is, as a relation between supply and demand. For example, carbon, which is found in all organic compounds, is essential to life. Yet, the economic value of carbon is computed "at the margin," for example, as the price of a lump of coal.

One might argue that species by sustaining ecosystems posssess economic or instrumental value. This may apply to prevalent species, such as worms, but no endangered species has been shown uniquely to support any ecological or other economic service. What such function, for example, does the American burying beetle, a listed species, perform? What ecosystem service on which we depend would collapse in its absence? As biologist David Ehrenfeld, the former editor of *Conservation Biology*, argues: Even a mighty dominant like the American chestnut, extending over half a continent, all but disappeared without bringing the eastern deciduous forest down with it. And if we turn to the invertebrates, the course of nearly all biological diversity, what biologist is willing to find a value—

conventional or ecological—for all 600,000-plus species of beetles?

The lack of market demand for all or virtually all endangered species in the United States confronts environmental economists with a dilemma. They could conclude that it is simply irrational for the nation to preserve endangered species. Yet, the Endangered Species Act is immensely popular. In surveys, Americans by large majorities agree with the statement that destroying species is wrong because God put them on this earth. Some commentators conclude. "It seems that divine creation is the closest concept American culture provides to express the sacredness of nature." The ESA appears, then, to reflect an obligation that Americans believe the nation ought to respect. This moral rule calls on us to preserve species for their own sake, not for the benefits they provide us.

Rather than declare species protection to be irrational, environmental economists have recognised the need to bring moral, spiritual, and other nonconsumption values into the welfare calculus. They have sought to show, therefore, that extinction is a kind of market failure. Environmental economists generally argue that markets fail to "Price" plants and animals correctly because "biodiversity is a public Good," From an economic perspective, the authors of the monumental *Global Biodiversity Assessment* declare, "the allocation of biological resources on the basis of market signals is inefficient." They conclude that "governments might intervene to correct the signals to private resources users."

The concept of a "public good" plainly does apply to some animals, such as tigers and whales, that people hunt in the wild. Economists quite properly recommend that private property regimes replace the rule of capture with respect to those organisms, such as salmon or halibut, that people seek to consume. Regulations that limit harvests by restricting access respond to the familiar problem of the commons.

Hunting and gathering wild plants and animals, however, is only an occasional cause of extinction these days. Endangered species are not generally threatened because humans consumer or directly use them. The problem is that humans, having no use for these creatures, wish to convert their habitats in economically profitable ways. Not hunting but habitat destruction or loss threatens the vast majority of endangered species.

Endangered species, moreover, like paintings, some of which hang in museums and some in private homes, often are found on property a government or an individual owns. The owners of property control access to these species, just as the owners of private and public galleries control access to art. As of 1993, about 80% of all species listed as endangered in the United States had some or all of their habitat on private land. Plants that grow on a person's land—flowers in one's garden, for example—seem to be examples of private not public goods. Even if species that reside on one's land are endangered, they seem to share the same properties of private goods—for example, rivalry and excludability in consumption—that belong to the ordinary flora and fauna found there.

Environmental economists generally reply that endangered species, even when found on private lands, are public goods because individuals are willing to pay to protect them, but this willingness to pay may be so dispersed that these individuals cannot make it effective by purchasing habitat. As a result, environmental economists have developed methodologies to assess "existence" and other values that express moral, cultural, and religious commitments to goals other than personal welfare or well-being—goals people associate with the good of society rather than their own good. Much current theoretical work in environmental economics concerns the measurement of these "non-use" values. This theoretical research has encouraged economists to bring into the welfare calculus the moral, aesthetic, and cultural commitments of Americans to goals other than welfare.

Over the last 20 years, resource economists have sought to find ways to bring into the social welfare calculus moral, religious, and other principled commitments to goals other than personal consumption. In the 1960s, environmental economists observed that individuals derive satisfaction when they forego material consumption to support policies they believe are intrinsically right. "There are many people who obtain satisfaction from the mere knowledge that part of wilderness North America remains." John Krutilla wrote, "Even though they would be appalled by the prospect of being exposed to it."

Building on Krutilla's theoretical insight, economists have developed many concepts—including "existence," "vicarious benefit," "bequest," and "stewardship" values—to capture in welfare terms amounts people are willing to pay for policies foreasons other than their own consumption opportunities. Much of the theoretical research in enviromnetal economics during the past decade has concerned the problem of assigning "shadow" (i.e., hypothetical market) prices to these moral beliefs or ethical principles. Contingent Valuation Methodology (CVM) uses surveys to elicit individual Willingness To Pay (WTP) for environmental pubic goods. This method, which involves direct questioning of respondents, has inspired a large literature responding to many technical problems of measurement that cannot be considered here.

The CV approach aappears most applicable to those

preferences that reflect self-interested goals, for example, an individual's desire to obtain a "warm glow" or sense of psychic satisfaction in knowing certain species exist. In these instances, a connection holds between the "existence' value of species and the welfare or well-being of individuals. Thus, insofar as economists are concerned about the welfare effects of environmental policies, their surveys should ask how much individuals would pay for the "warm glow," or psychic income, or other personal benefit associated with environmental protection. As a rule, however, these surveys inquire how much people are willing to pay simply to protect species or to secure other nonconsumption goods. As long as people care abut environmental goals for reasons besides their own advantage, however, the survey results would have nuclear relation to their welfare or well-being.

Empirical research has found that "responses to CV questions concerning Environmental preservation are dominated by citizen judgements concerning social goals and responsibilities rather than by consumer preferences." Responses to CV surveys reveal "social or political judgements rather than preferences over consumer bundles." A careful study by two economists concurred that ethical commitments often dominate economic or welfare considerations in responses to CV surveys. "Our results provide an assessment of the frequency and seriousness of these noneconomic considerations. They are frequent and they are significant determinants of WTP responses."

That a person's desires "are all his desires," as philosopher Mark Overold has written, "does not seem sufficient for saying that satisfying one or all of them would promote his welfare." On the contrary, the principled beliefs on the basis of which individuals form environmental values may reflect spiritual, religious, moral, political, cultural, and other commitments that bear no clear connection to benefits individuals may expect themselves. Critics have concluded that "existence' and other non-consumption values are unrelated to personal welfare or well being and thus to the welfare goals of environmental economics.

As philosopher Ronald Dworkin has pointed out, most Americans attribute intrinsic value to natural objects such as other species. We "think we should admire and protect them because they are important in themselves, and not because we or others want or enjoy them." Economist Amartya Sen has observed that our commitment to moral values drives a "wedge between personal choice and personal welfare," although "traditional economic theory relies on the identity of the two."

## THE RELATION BETWEEN WILLINGNESS TO PAY AND WELL-BEING

If distributive or equity goals are considered separately, environmental economists believe that human welfare or well-being is the only object that has intrinsic value for the purposes of public policy. (Economists such as Burton Weisbrod have pointed out that equity concerns are also preferences for which people are willing to pay and so may be folded into the cost-benefit calculus on the same basis as any other good). When one considers what mainstream economists mean by "welfare"—the satisfaction of preferences taken as they come bounded by indifference and ranked by willingness to pay—one may ask why it provides a normative basis (indeed, the only normative basis) of environmental policy. The term "welfare," as used by economists, has no connection, either logical or empirical, with concepts such as contentment or happiness. Having a preference (for example, to smoke a cigarette) may give the individual who has it a reason to try to satisfy it. Why has society a reason, however, to satisfy these personal or private desires?

The central normative postulate of environmental economics asserts that preferences should direct allocation. The postulate rests on the assumption that insofar as a person's preferences are satisfied, his or her well-being is increased. The term "satisfy" is used in this context in the mathematical sense in which equations or conditions are "satisfied." The unexamined postulate of welfare economics is that to "satisfy" a preference in this *mathematical* sense is also to satisfy a person in the *psychological* sense of making him or her happier or more content.

Empirical studies show, however, that once people meet their basic needs, they do not become happier or more content as then income increases and therefore they are more able to satisfy their preferences. Social research confirms what common wisdom suggests: Money and with it preference satisfaction do not buy happiness. One review concludes, "Studies of satisfaction and changing economic conditions have found overall no stable relationship at all." Those "who win large sums of money in football pools or lotteries are not found to be on the whole more happy afterwards."

Contentment depends more on the quality of one's desires and on one's ability to overcome them, than on the extent to which they are satisfied. The things that make one happy—friends, family, achievement, health—depend on virtue and luck; they are not for sale. Consumption does not produce contentment. A philosopher explains, "this is virtually inevitable because the faster preferences actually are met, the faster they escalate."

A legal theorist correctly notes that the "most important thing to bear in mind about the concept of value [in the welfare economist's sense] is that it is based on what people are willing to pay for something rather than the happiness they would derive from having it." If social welfare is defined as a function of "what people are willing to pay for something" rather than "the happiness they would derive from having it," it is not surprising that a policy that allocates resources to those willing to pay the most will maximise welfare. The goal of maximising well-being—if "well-being" is merely a function of willingness-to-pay—merely restates but does not justify the goal of allocating resources to those willing to pay the most for them.

Welfare economists, in other words, offer no method for measuring changes in well-being other than willingness-to-pay. Thus, for them, the relation between willingness-to-pay and improvements in well-being is a stipulative or definitional rather than a contingent or testable one. Accordingly, "welfare" is just another name—a proxy or a stand-in—for "efficiency." By allocating resources to satisfy the preferences of those who are willing to pay the most for them, society will, of course, maximise social welfare, because "social welfare" is itself defined in terms of the satisfaction of those preferences.

If one could assume that the individual always prefers what he thinks will increase his or her well-being, then the relation between well-being and preference might be more than a stipulative one. But surveys have shown again and again that people value objects and outcomes for many reasons besides the benefits they offer them. The question then arises if these reasons—for example, a moral obligation to protect individuals from the "trespass" of pollution or to prevent extinction—offer as important grounds for policy as the satisfaction of "given" preferences on the basis of willingness to pay. The ESA, the Clean Air Act, and other environmental statutes apparently base policy on deontological rather than utilitarian grounds—on principles rather than on preferences. For the environmental economist, however, the goals of public must be utilitarian for otherwise they would be irrational. If we take this scientific position, we must base public policy not on the moral principles implicit in public law but on the given data of private preferences. The objective description of these data is at present the principal preoccupation of the science of environmental economics.

One of the most difficult technical problems confronting environmental economists is that of identifying the preferences that motivate behaviour. This problem arises because preferences, being private psychological states, are not directly observable. Indeed, of all the factors that determine action—long term goals, habits, beliefs, mistakes, impulses, obligations, hopes, fears, and so on—preferences may not figure prominently. Preference may best be understood as a theoretical construct of welfare economics inferred from descriptions of behaviour, not as a mental entity otherwise of interest to psychologists.

Economists infer preferences from descriptions of behaviour. Behaviour, however, does not describe itself; indeed, the bodily actions of an individual can be interpreted in any number of ways. Every object one might choose, moreover, possesses an indefinite number of qualities, any one of which might motivate that choice. Everything we do may be explained in the context of a thousand impinging concerns unique to the person and the situation. It would be hard for the individual to reconstruct the alternatives he or she chose among and thus to describe the exact choice he or she made. Welfare and environmental economists circumvent this problem by stipulating the "opportunity set" and by describing the choice and then inferring a preference from that deception. But choices are rarely if ever transparent or obvious, and even the greatest psychoanalysts have despaired at plumbing the soul.

From the perspective of applied ethics, the problem

with the normative policy sciences is that there are so many of them. Libertarians, Marxists, Utilitarians, Egalitarians, Kantians, and others propose "foundations" for various areas of the law; they disagree not only with each other but also with legal realists, economic determinists, and environmental catastrophists. The deep ecologist who sees the earth as a organism thinks he or she is just as objective—just as scientific—as the economist who speaks in terms of commodities. There may be as many objective and scientific views of any policy question as there are opinions, since nobody believes his or her view is only a subjective preference.

Because so many scientific positions contend for political vindication, as philosopher John Rawls has argued, the conception of justice in a constitutional democracy "should be, so far as possible, independent of controversial philosophical and religious doctrines." Democratic political processes provide the best means we know to work out controversies about the philosophical basis of public policy. Welfare and environmental economics presents one controversial doctrine about the appropriate goals of public policy. The goal of maximising the satisfaction of preferences taken as they come and as determined by experts may be an important one. It must be defended, however, against other approaches to environmental policy, for example, deontological views of pollution as an enjoinable tort and extinction as a kind of crime. These alternative views of public policy have carried the day in Congress. They are based on moral and community-regarding judgements at least as worthy of consideration on their merits as are the opinions of environmental economists. Instead, economists too often dismiss these opposing views as weird kinds of nonconsumption consumer preferences that markets have failed to price and for which they must therefore develop techniques of elicitation and measurement.

## ENVIRONMENTAL COMPLIANCE BY INDUSTRY

It considers the extent to which organisations act in accordance with standards established by government to protect the environment. The notion of ecologically sustainable development is being employed to encourage organisational to be mindful of the need to consider ecological as well as economic factors when developing organisational activities. There is a range of instruments available to influence the attitudes and behaviour of people in industry toward environmental and ecological issues. The range extends through various "command and control" instruments, market-based measures, and self-regulatory devices. Command and control instruments require little in the way of ethical judgement, apart from the decision of an organisation whether to comply with mandated behaviour given the associated penalty and reward structure. On the other hand, at the latter end of the spectrum ethical issues have a role at two states—the decision to undetake voluntary regulation, and the decision to comply with the norms of self-regulation once introduced. This article explores available instruments, the motivation to comply with each type of instrument, and incentive structures including the monitoring and reporting of compliance and noncompliance by industry.

## SUSTAINABLE DEVELOPMENT, AND ENVIRONMENTAL COMPLIANCE

"Industry's response to pollution and resource degradation has not been and should not be limited to compliance with regulations. It should accept a broad sense of social responsibility and ensure and awareness of environmental considerations at all levels".

Development of industry has made it possible for us to increase the number of human beings the world can support. At the same time, environmental problems have accompanied industrial development. Until the 1980s industrial development and the environment were looked at as separate issues. The concept that has brought them together is Ecologically Sustainable Development (ESD). This concept is attemping to get industry to perceive the environment as an asset—one to be nurtured, one to maintain, one to invest in, one to protect for present and future generations and promote intergenerational equity. Indeed, a commonly quoted definition of sustainable development, as recorded in the Bruntland Report 1987, is "development that meets the needs of the present without compromising the ability of future generations to meet their own needs".

In order to establish whether progress is being made in attempts to achieve ESD processes in industry, there is a need, first, to measure the adverse impact an organisation is imposing on the nature environment; second, there is a need to disclose information about the environmental impacts of each organisation to its stockholders; and third, there is a need to assess the extent to which organisations have attempted to achieve the environmental standards set down for them. One set of standards is established through legislation. Compliance with legal standards has received much attention, but many organisations go "beyond compliance" with basis legal requirements. The pertinent ethical question is how far should we expect an organisation to feel morally responsible for voluntary self-analysis with regard to the impact of its activities on the well-being of our environment? Against a background of differing legislation, is it justifiable for multinational organisations to undetake the same operations in developed and developing countries while varying the environmental standards they adopt?

Crucial monitoring, investigating, and reporting of environmental standards achieved by industry has really only begun in earnest the last decade. It is hampered by cost, by creative disclosure, and by the sheer volume and diversity of industrial activities that have an impact on the environment. These range from issues of biodiversity and air and water pollution to land degradation. Progress tends to have been implemented in an incremental way with emphasis initially being focused on identifying industry sectors that are thought to have a significant impact on the physical environment. These are shown in Table11.1

**Table 11.1**
**Industry Sectors thought to have a Significant Impact on the Environment**

| *Industry* | *Environmental Problems* |
|---|---|
| Agriculture | Land digression; soil erosion; desertification; salinity and acidification. |
| Energy production and use | Global warming and air pollution caused by use of fossil fuels |
| Fisheries | Depleting stocks; endangered species |
| Forest use | Logging of old growth forests; water and air pollution from pulp and paper manufacture; rain forest depletion. |
| Manufacturing | Air, water and land pollution |
| Mining | Depletion of natural resources; land degradation at mine sites; global warming; water pollution from open cast tailings |
| Tourism | Destruction of wilderness |
| Transport | Air pollution; global warming |

Three particular concerns relate to industries:

* where the environment is being destroyed through irreversible processes—such as the wholesale destruction of tropical rain forest and the loss of biodiversity;

* Where environmental problems are common to a number of different industrial sectors—such as the greenhouse effect of "global warming";
* Where sustainable development is a global rather than a local issue—such as transboundary pollution caused by industrial waste fouling the waters of the Rhine.

Solutions are being explored at local, national, and international levels.

## ENVIRONMENTAL STANDARDS FOR INDUSTRY

Environmental standards with which individual organisation may or may not comply can be established in six main ways:

* directly by government through legislation;
* indirectly by government through the introduction of economic and administrative incentives designed to encourage acceptable environmental behaviour;
* through pressure from community interest groups;
* through third parties that bring pressure to bear on organisations;
* through voluntary self-regulation by industry associations; and

through self-regulation by individual organisations.

### 1. Environmental Legislation

At federal, state, and local levels of government many thousands of environmental laws have been implemented in recent years. For example, at the federal level in the United States, there are more than 11 thousand environmental laws. Sovereign states are also under pressure to ratify international environmental conventions, thereby adding to the volume of legislation.

Environmental legislation has traditionally been the method used by governments to establish standards with which organisations much comply if they are to avoid penalities. Organisations are aware that they are only allowed to exist because governments recognise that they provide something of value to society. Hence, most organisations respond favourably to required behaviour mandated by legislation. To encourage compliance, legislation usually specifies penalities for noncompliance. This represents a "command and control" approach to achieving environmental compliance. Penalties may be imposed for any form of behaviour toward the environment that is unacceptable to society as reflected in legislation. Unacceptable behaviour includes such actions as excessive pollution of the air, destroying the potability of water, and degrading soils with toxic substances. Penalities range in form from simple monetary fines imposed on organisations, or individuals such as company directors who are responsible for organisational activities, to time spent in jail for company officials who are found guilty of the crime of degrading the environment in a number of specified ways. One problem with using legislation to achieve environmental compliance by industry is that organisations tend to try and achieve the environmental standards laid down, but they do not attempt to improve performance beyond this legally required minimum. Legislation does not encourage an ethical commitment to go beyond legislated standards of behaviour. Neither does it promote the ethos of a constant search for improvement in environmental performance.

Finally, complete compliance with legislation is not assured without a moral commitment by the organisations affected. In the absence of a police state, it is not possible to attempt to monitor the environmental impacts of all organisational activities at all times. As controls proliferate, the costs to government of monitoring begin to out weight the benefits of protecting the environment. Hence, emphasis tends to be placed on legislation that is designed to ensure compliance by organisations that are easy to target, such as large companies rather than small business, and environmental problems that are easy to focus on, such as point-source pollution of waterways by chemical companies rather than broach run-off water pollution caused by agricultural practices. In summary, while direct environmental regulations may be effective in achieving the goals established, legislated solutions can be inflexible and can impose high administrative costs of compliance on the community.

## 2. Economic and Administrative Incentives

### *(a) Economic Incentives*

One way for government to encourage a moral commitment to consider environmental impacts when organisations make decisions is to provide incentives for good Environmental behaviour rather than penalties for non-compliance with legislation. There are two main types of incentives to consider: economic and administrative (see Table 11.2).

Economic instruments are designed to induce rather than direct compliant behaviour. Their appeal lies in a shift of responsibility from the regulator to the user, and a focus on the efficient and innovative use of resources to meet environmental quality targets. While legislation is aimed at directly changing the behaviour of polluters by fiat, economic instruments aim to make environmentally damaging bahaviour cost more.

**Table 11.2**
**Typical Penalties and Incentives Designed to Protect the Environment**

| *Penalties* | *Incentives* |
|---|---|
| *Economic* | *Economic* |
| Direct user charges | Direct grants and subsidies |
| Emission and effluent charges | Bounties |
| Licensing fees | Commissions |
| Environmental taxes | Environmental tax credits |
| Tradeable quotas | Tradeable emission rights |
| Performance bonds and guarantees | Deposit refunds |
| Proportional noncompliance fees | |
| *Administrative* | *Administrative* |
| Tighter monitoring | Favourable administrative consideration—ease monitoring of compliant organisations. |
| Poor Publicity | Prizes and rewards as symbols of good behaviour—ecolabelling, and certificates for the voluntary introduction of environmental management systems. |
| Condemnation | Praise |

Economic instruments engage only when an organisation transacts in a market. In some cases markets have to be established in order to help solve environmental pollution. Such markets create rights to use environmental resources, or to pollute the environment up to a predetermined limit, and they allow these property rights to be traded. One example is tradeable pollution rights. These rights permit factory emissions into the atmosphere. If an organisation owns some of these rights to pollute, but it does not need to pollute up to the maximum permitted because of say, the introduction of environmentally benign technology, the rights can be sold to another organisation that wishes to buy the right to pollute the air. Such an organisation might be a new business that is trying to establish itself in a region where the total volume of a particular air pollutant (such as carbon dioxide emissions) has been established and the rights to pollute issued. In practice, the United States is the only country where emissions trading is widely applied, and the aim is to achieve economic efficiency rather than environmental effectiveness. In other words, the aim is to reach a predetermined air quality standards at the least cost to industry, and to promote growth despite restrictions on increasing air pollution.

A number of problems exist with tradeable pollution rights. These include; high administrative costs of establishing and trading in the market; difficulty in establishing the assimilative capacity of the atmosphere for different pollutants; thin trading in pollution rights making quoted prices unrealistic; and the moral problems associated with tracing the right to pollute the air as a commodity to be bought and sold for profit when it was previously considered as a common good to be maintained in good condition and to be freely used by all biota.

Opposition to emissions trading suggests that it encourages the view that polluters should decide whether there should be a trade-off between economic development and environmental quality, rather than regulators who represent individuals with inalienable rights to clean air.

Where property rights in markets are already established, market prices can be adjusted to bring about a more desirable outcome than an unfettered market would achieve on its won. Price-based measures are used to achieve optimal levels of environmental damage by inernalising environmental costs in organisations that would otherwise have passed these costs on to third parties by treating environmental resources as a free good.

It is argued that if polluting industries pay for the full cost of the environmental damage they cause, then society is no worse off because the payment can be used to rectify any environmental damage. There are several problems with this argument. First, money obtained from pollution charges is rarely used to correct environmental damage; rather, it is used as general revenue by governments. Even if it was so used it is not clear that the environment could be restored. The case of ozone depletion is illustrative. Certain chemicals released to the atmosphere by industry track into the stratosphere and deplete ozone. Destruction of the ozone layer may threaten the existence of biota. Should industry be allowed to buy pollution rights to release ozone depleting chemicals, such as halons, into the atmosphere? Concern in the world community about the irreversibility of ozone depletion led to the decision that this was not a gas that could be released into the atmosphere at any price. Production and use were banned through international convention, national legislation, and a willingness by industry to find replacement chemicals.

Second, the complitation of the environmental costs for which compensation is to be sought is difficult. Environmental accountants are slowly beginning to address this issue. Third, in ethical terms, selling rights to pollute today stands in opposition to the notion of humanity acting as a stewad of nature for the benefit of future generations. Likewise, the need for intergenerational equity is one reason why it is argued that future environmental quality should not be determined by those who currently possess the financial resources to buy a current right to pollute. Fourth, the use of economic incentives encourages an anthropocentric rather than an ecocentric approach to receiving environmental problems. Anthropocentric views treat humans as the centre focus of analysis, whereas ecocentric views treat human beings as only one part of an ecological system. To an ecocentric, what might be justified from an anthropocentric perspective may have no moral justification. For example, from an anthropocentric view trade in endangered species is to be encouraged using economic incentives to protect and increase numbers of species that will otherwise disappear. From an ecocentric view, it is considered unethical to determine the fate of a species based on the amount of profit that can be made from trade. Trade in endangered species is regarded as akin to trade in human slaves.

### (b) Administrative Incentives

Administrative instruments provide an alternative way of getting organisations to act in a being way towards the environment. They also rely on the market, but they act to enhance the reputation and the legitimacy of organisations in the view of people operating in the market. Hence, good reputation and increased legitimacy in environmental affairs can command a price premium for an organisations, through the moral involvement of shareholders, customers, and employees. A price premium may also be obtained from government as a good environmental performer may have the administrative costs of monitoring by regulators eased. Governments have considerable discretion in the extent to which they enforce the requirements of environmental legislation.

There is a range of administrative incentives available to encourage compliance with government environmental policy. First, as a reward for organisations that are proactive in their consideration of environmental impacts, local and national governments, and international bodies, may reduce the administrative burden of compliance. When an organisation has developed a record of good environmental behaviour, and has taken effective action to rectify any problems that have come to light, it may be appropriate to demonstrate that a degree of trust exists between the regulator and the regulated. Recognition could be given by reducing the extent and periodicity of monitoring of organisations that have an established record as good performers, or as performers that have gone beyond the legislated standards, or that are industry leaders. This would help reduce their costs of compliance and it would provide a competitive advantage over organisations that have a record of noncompliance. The problem with this approach is that the system might be open to abuse. Enforcement agencies could be subject to pressure to provide exemptions when they were not merited.

Second, excellence in the presentation and disclosure of inforamtion about an organisation's environmental performance could be rewarded by the award of a prize and the associated good publicity that would accompany a presentation. For example, in the United Kingdom an annual award is made to the organisation publishing the best corporate environmental report. In 1995, the award, made by a professional accounting body, was presented to Thorn EMI. Thron EMI was praised for presenting a life-cycle assessment of each segment of its business showing the environmental impacts at each stage of the life cycle; for including physical measures of environmental targets and an assessment of the extent to which targets were achieved; for including information about the use of ozone depleting substances; and for reporting separate categories of environmental expenditure during the year. Although there is no clear direct economic benefit to an organisation from gaining an award, indirect benefits include improved standing in the community and improved employee motivation, and disclosures reduce the risk that unidentified environmental liabilities might emerge to reduce the wealth of shareholders. Certificates awarded for voluntary implementation of audited internal environmental management systems, and logos for display on environmentally benign products, have similar symbolic effects. Both provide motivation for organisations to iternalise the environmental management challenge and to go beyond a strategy of grudging compliance with government legislative requirements.

## 3. Community Interest Groups

Community interest groups, such as environmental Non-Government Organisations (NGOs), play an increasingly important role in bringing pressure on industry to comply with environmental legislation. The reason is that government and industry in many countries respond to public opinion, and public opinion about the environmental record of companies is molded by NGOs. They take direct action to raise awareness about suspected environmental abuses. For example, in 1995 Greenpeace waged a campaign against the Shell Petroleum Company's decision to sink an oil rig in the North Sea. The main impact of Greenpeace was to encourage consumers not to buy Shell's petroleum products, particularly in Europe. The effect was such that eventually Shell was obliged to reverse its decision and it agreed to break the rig up on land. NGOs also attempt to educate the public about poor environmental practices in industry. In the process, they bring moral suasion to bear on industry.

## 4. Third Parties

Banks and insurance companies are also bringing overt pressure on organisations to company with environmental legislation and to introduce environmental management systems that can identify environmental risks. If property offered as security for a loan has been polluted by toxic waste, the client might go bankrupt. In this case, the bank takes over as mortgagee. In some countries, for example in the United States under the "Superfund" legislation, the bank then could become responsible for the cost of cleaning up the contamination. To avoid this possibility, before they lend money to industry, banks now insist upon an invironmental audit of property offered as security by a prospective mortgagee. Similar precautions are taken by banks when they lend to organisations that are involved in takeovers and amalgamations.

Insurance companies provide another example of third-party pressure for improved environmental performance. Before, they will issue insurance cover against environmental risks they need to receive evidence that an environmental audit has been undertaken.

## 5. Codes of Environmental Management Practice and Industry Associations

In the last decade, codes of environmental management practice have emerged as a third method of establishing environmental standards and monitoring compliance by organisations. General and industry specific codes dominate. Several countries in the developed world encourage the voluntary adoption of standards set for environmental management by the International Standards Organisation (ISO). Best practice environmental management systems are outlined in the ISO14000 series of standards. These were first introduced in Britain as the British standard BS 7750. Examples of a general code include the International Chamber of Commerce's (ICC) Business Charter for Sustainable Development (see Table 11.3), and the principles of the Coalition for Environmentally Responsible Economies (CERES), which came into being following the grounding of the Exxon Valdez and the consequent spill of oil in Prince William Sound.

## Table 11.3
## International Chamber of Commerce and the Business Charter for Sustainable Development

*Principles*

1. Corporate priority.

   To recognise environmental management as among the highest corporate policies and as a key determinant to sustainable development; to establish priorities, programmes, and practices for conducting operations in an environmentally sound manner.

2. Integrated management.

   To integrate these policies, programmes and practices fully into each business as an essential element of management in all its functions.

3. Process of improvement.

   To continue to improve corporate policies, programmes, and environmental performance, taking into account technical developments, scientific understanding, consumer needs, and community expectations, with legal regulations as a starting point; and to apply the same environmental criteria internationally.

4. Employee education.

   To educate, train, and motivate employees to conduct their activities in an environmentally responsible manner.

5. Prior assessment.

   To assess environmental impacts before starting a new activity or project and before decommissioning a facility or leaving a site.

6. Products and services.

   To develop and provide products or services that have no undue environmental impact and are safe in their intended use, that are efficient in their consumption of energy and natural resources,and that can be recycled, reused, or disposed of safely.

7. Customer advice.

   To advise, and where relevant educate, customers, distributors and the public in the safe use, transportation, storage, and disposal of products provided; and to apply similar considerations to the provision of services.

8. Facilities and operations.

   To develop, design, and operate facilities and to conduct activities that take into consideration the efficient use of energy and materials, the sustainable use of renewable resources, the minimisation of adverse environmental impact and waste generation, and the safe and responsible disposal of residual wastes.

9. Research.

   To conduct or support research on the environmental impacts of raw materials, products, processes, emissions, and wastes associated with the enterprises and on the means of minimising such adverse impacts.

10. Precautionary approach.

    To modify the manufacture, marketing or use of products or services, or the conduct of activities, consistent with scientific and technical underestanding, to prevent serious or irreversible environmental degradation.

11. Contractors and suppliers.

    To promote the adoption of these principles by contractors acting on behalf of the enterprise, encouraging and where appropriate, requiring improvements in their practices to make them consistent with those of the enterprise; and to encourage the wider adoption of these principles by suppliers.

12. Emergency preparedness.

    To develop and maintain,where significant hazards exist, emergency preparedness plans in conjunction with the emergency services, relevant authorities, and the local community, recognising potential transboundary impacts.

13. Transfer of technology.

    To contribute to the transfer of environmentally sound technology and management methods throughout the industrial and public sectors.

14. Contributing to the common effort.

    To contribute to the development of public policy and to business, governmental, and intergovernmental programmes and educational initiatives that will enhance environmental awareness and protection.

15. Openness to concerns.

    To foster openness and dialogue with employees and the public, anticipating and responding to their concerns about the potential haards and impact of operations, products, wastes, or services, including those of transboundary or global significance.

16. Compliance and reporting.

    To measure environmental performance; to conduct regular environmental audits and assessment of compliance with company requirements, legal requirements, and these principles; and periodically to provide appropriate information to the Board of Directors, shareholders, employees, the authorities, and the public.

---

*Source:* Willums, J.O., and Goluke, U. (1992). *From ideas to action: Business and sustainable development.* Oslo: International Environmentâl Bureau of the International Chamber of Commerce.

The packaging industry's voluntary Environmental Code of Practice for the post-production impact of new packaging materials provides a case in point (see Table 11.4). In the Code, functions of packaging are identified. These functions include ensuring that products are protected and preserved, waste is minimised, quality, health and safety are assured, retail self-service is promoted, and that stocks are secure prior to sale. But the Code also recognises that the Environmental impact of some new forms of packaging is hard to gauge before their introduction into the marketplace. The bodies responsible for establishing the Code see adherence to it, both literally and in spirit, as applicable to all members of the packing and filling industry. In short, an ethical commitment and a moral responsibility to the new Code, which acts as a supplement to legislation already covering exiting packaging materials, is sought from members. Enforcement of the Code resides with an advisory council, which is charged with the task of promoting the Code to industry and developing, assessing, and reviewing implementation measures considered necessary to ensure conformance with the Code by industry. In addition, if a new package does not conform to the code, the council will work with the manufacturer to improve the package on a case-by-base basis.

**Table 11.4**
**Environmental Design Considerations for New Packaging**

| | *General Principles* |
|---|---|
| 1 | The package should be designed with due regard for its ultimate disposal in order to minimise its impact on the environment. |
| 2 | The waste management options which should be considered are: reduction, re-use, recycling, disposal. |
| 3 | It may be useful to include consumer education relating to the package in promotion and marketing materials. |
| 4 | If the package is to be exported it should be determined whether a special export pack is required to satisfy environmental and other requirements of the export market. |
| 5 | Consideration as to how the package is to be distributed and whether distribution requires special conditions such as refrigeration should be determined. The impact of the proposed distribution system on the environment and the energy usage of the system should be taken into consideration. |

*Source:* The Packaging and Filling Industry. (1992). *Environmental code of practice for the packaging industry,* pp. 11-12.

The chemical industry's Responsible Care Programme and its accompanying Code of Practice represent the most sophisticated, advanced, and potentially effective scheme of self-regulation of environmental performance by an industry. Chemical industries in more than 20 countries have established a Responsible Care Programme. Anxiety and distrust of the chemical industry, a series of major environmental disasters throughout the world, and the threat of greater government involvement in regulation combined to promote the development of principles that act as statements of intention. These principles can be found in Table 11.5. Individual organisations evaluate their own performance against these principles, and their reports are collected and used to tell the public how well the industry is doing Moral suasion, or peer pressure, is

**Table 11.5**
**The Responsible Care Programme in the Chemical Industry: Statement of Intention**

| | *Abiding principles* |
|---|---|
| 1. | To recognise and respond to community concerns about chemicals and our operations. |
| 2. | To operate our plants and facilities in a manner that preserves the environment and protects the health and safety of our employees and the public. |
| 3. | To develop and produce chemicals that can be manufactured, transported, used, and disposed of safely. |
| 4. | To give health, safety, and environmental considerations priority in our planning for new products and processes. |
| 5. | To report information on relevant chemical-related health or environmental hazards promptly to appropriate authorities, employees, customers, and the public, and to recommend protective measures. |
| 6. | To give advice to customers on the safe use, transportation, and disposal of chemical products. |
| 7. | To increase knowledge by conducting and/or supporting research on the health, safety, and environmental effects of our products, processes, and waste materials. |
| 8. | To cooperate with customers, authorities, and affected groups and individuals to resolve problems created by the handling and disposal of chemical substances considered hazardous. |
| 9. | To cooperate with government in developing laws and regulations to safeguard the community, the work place, and the environment; to endeavour to ensure that such laws are based on scientifically supported data and/or expert opinion. |
| 10. | To promote these principles and practices by sharing experience and offering assistance to others who produce, handle, use, transport, or dispose of chemicals. |

*Source:* Beder, S. (1993). *The nature of sustainable development*, p. 237, Newham, Australia: Scribe Publications.

used to encourage chemical companies to remain within the programme. Lack of compliance with the guidelines can lead to expulsion from the chemical industry association. The importance of the Responsible Care Programme is that it is an attempt to introduce industry self-regulation in order to preempt further government regulation.

## 6. Self-Regulation

Growing concern over environmental problems is leading organisations to consider the acceptance of a moral duty to protect the environment. Government legislation, thrid-party influences, and codes of practice offering guides to benign environmental action all bring pressure on organisations to go beyond compliance with the minimum standards and codes. Ultimately, however, many environmental problems will not be solved without a moral commitment from the individuals who are responsible for organisational activities.

Management philosophies that consider an organisation's impact on the environment are being encouraged. Mention has already been made of the ISO 14000 series, which addresses environmental management systems (EMS). In addition, the present popularity of the philosophy of Total Quality Management (TQM) places emphasis on the need for continuous improvement in process, products, and services, looks toward the involvement of suppliers and customers in the processes of continuous improvement, and reduces the emphasis on inspection-based quality controls.

Adoption of these systems is purely voluntary. Some of the reasons they might be adopted include:

* Organisations adopt environmental management systems in order to control and promote their own activities rather than have them regulated by government;

* Organisations perceive that they can make financial gains by building up an image of being environmentally concerned;

* Voluntary environmental disclosure may lead to improved legitimacy of the organisation, especially if reported information puts the organisation in a good light;

* The direct costs of collecting and collating environmental information are falling as improvements in information technology facilitate the development of systems to measure and record environmental data; and

* The direct costs of reporting environmental information are falling as experience increases. Experimentation has begun with the disclosure of environmental performance data within annual reports and in separate environmental reports. For example, over a number of years, The Body Shop International has produced environmental reports on all aspects of its activities and has published them in *The Green Book.* Thorn EMI, a British company that won a prize for the best environmental report in 1995, includes a summary 4-page environmental report in its annual report and accounts targeted towards investors. The information is gleaned from a separate 36-page environmental report available for the specialist environmentalist.

Some organisations are unlikely to adopt sound environmental management principles voluntarily because:

* Their organisation is having financial difficulties, and cannot make the necessary investment in environmentally benign technology;

* Their organisation is too small to muster the resources, including specialised skilled personnel.

In practice, a combination of these methods is used to achieve the goals of environmental policy. The choice of this combination in any particular country is referred to as the "regulatory mix." For example , when comparing legislation with economic instruments the advantages and disadvantages need to be balanced on a case-by-case basis. An appropriate combination of compliance techniques may sometimes be more effective (see Table 11.6)

**Table 11.6**

**The Advantages and Disadvantages of Legislation and Economic Incentives**

| *Legislation* | *Economic instruments* |
|---|---|
| *Advantages* | |
| • They determine the objectives and means independently of economic factors and market forces. | • They allow each polluter to choose the most effective and efficient way of reducing pollution. |
| • They are the surest means of preventing irreversible effects or a totally unacceptable level of pollution. | • They provide an incentive to continue reducing pollution in order to reduce costs. |
| • Their objectives, criteria, and surveillance measures are widely understood. | • They minimise the cost of achieving pollution reductions. |
| | • They provide finance for other things including restoration of damage. |
| *Disadvantages* | |
| • They may not be the most cost-effective way of ensuring environmental standards are met. | • They depend on trial and error to get the charges for environmental degradation right. |
| • They do not provide an incentive for polluters to do better than standard. | • Setting up arrangements for supervising and measuring is expensive. |
| • They are generally difficult to administer, and enforcement of standards depends on the resources available to the regulatory authority. | |

*Source:* House of Representatives Standing Committee on Environment and Conservation (HRSCEC). (1987). *Fiscal measures and the achievement of environmental objectives,* pp. 14-15, Canberra, Australia: AGPS.

## PROBLEMS OF MONITORING

No recognition of compliance by industry can be established unless adequate monitoring, reporting, and inspection takes place. Whatever methods are exacted on industry—legislation, incentives based mechanisms, or self-regulation—monitoring and reporting of environmental performance is crucial. Because it is time-consuming and expensive for governments to monitor each organisation directly, and because there has been a change in perceptions about the appropriate scope and means of government intervention in general, indirect and voluntary means have been encouraged in recent years. Successful self-monitoring of the extent of compliance depends upon periodic reports about the environmental performance of each organisation. This forms one element of an environmental compliance programme.

An environmental compliance programme is an organised, systematic approach to ensure that an organisation is in compliance with all applicable environmental laws, regulations, and standards. The purpose of a compliance programme is to minimise the risk that an organisation will have to spend time and resources dealing with environmental problems at some point in the future, or will receive adverse publicity from an environmental disaster caused by the organisation. By drawing attention to possible environmental problems, it permits managers to act in advance to remove environmental risks to the community, to employees, and to the organisation. The concept of ESD gives explicit support to this approach by invoking the proactive "precautionary principle" in environmental management.

An environmental compliance programme is necessary when an organisation:

- Faces significant environmental risk because of the nature of its industry;

* Has many permits and needs to monitor compiliance with the permits where there are administrative benefits to be gained from a good record, or penalties to be incurred in the event of noncompliance;
* Has a poor historical record in relation to its management to environmental matters;
* Has been the subject of previous civil or criminal enforcement actions;
* Has ethical investors who invest only if environmentally responsible behaviour can be demonstrated;
* Is anticipating acquisition activity, or the need to undertake insurance against environmental risks.

The main elements of an environmental compliance programme include:

* A set of compliance policies that are tailor-made for the organisation;
* Compliance audit procedures that are independent of operations, and are designed to identify and correct areas of noncompliance;
* Internal and external reporting systems designed to encourage continuous improvement and innovation in environmental matters;
* A formal identification of who is responsible for organising a response to environmental emergencies, and who is responsible for provision of "right-to-know" information to the community;
* Education and training for staff responsible for possible environmental mishaps;
* A mechanism for identification, measurement, and disclosure of both positive and negative environmental information;

* Incentive schemes that encourage improved environmental performance; and
* Procedures for classifying environmental expenditures and a mechanism for linking these to responsible individuals or groups.

The key component of an environmental compliance programme is a suitable auditing, monitoring, reporting, and tracking system. Demonstration of a successful compliance programme can be used as evidence in a defense against any action brought for noncompliance.

One way an organisation can demonstrate that it has complied with legislated environmental standards is to publish a Compliance with Standards (CWS) Report, either as part of its annual accounts or in a special environmental report. A CWS Report demonstrates the extent to which an organisation has satisfied the various elements of a compliance programme.

As the regulatory mix is gradually extended, ethical issues are growing in significance in the calculus designed to ensure environmental compliance by industry. The global trend in the developed word toward a small public sector means that greater reliance will be placed in the future on voluntary conduct by national and multinational organisations. Voluntary compliance with environmental norms established as a basis for industry performance is likely to continue to be driven by financial incentives as well as by ethical considerations for community welfare.

## JUSTICE TO ENVIRONMENT

Environmental matters are external and prior to issues of social justice, arguing instead that social and economic relations of class, race, and gender are central to how environmental issues are identified and how they are fought. The differences in power and opportunity that discrimination on the grounds of race, class, and gender create mean that some sections of the population suffer a disproportionate burden of the detrimental effects of environmental degradation. Environmental justice recognises the connection between environmental and human well-being and the organisation of production in society, whose present relations of exploitation and inequality are directed at the continuation of privilege for some sections of society.

## THE EVERYDAY ENVIRONMENT

The environmental justice movement emerged in the United States in the 1980s as a development of the movement for social justice which sought to challenge racial, sexual, and economic discrimination. The movement strove to politically empower communities so as to include them in decisions that were to affect them, and campaigned to improve home and workplace conditions.

Increasingly the concerns of this community-based movement came to be centred around environmental issues of public health, occupational safety, and, in particular, the siting of polluting industries and technologies and waste disposal facilities. This was especially the case in communities of colour which were (and remain) disproportionately exposed to hazards of these sorts. Emerging from these urban, community-based concerns came a movement for which the term "environment" took on a radically new meaning. As Dana Alston, speaking at the first national People on Colour Environmental Leadership Summit in1991, stated, the environment, for us, is where we live, where we work, and where we play. The environment affords us the platform to address critical issues of our time: questions of militarism and defense policy; religious freedom; cultural survival: energy-sustainable development; the future of our cities;

transportation; housing; land and sovereignty rights; self-determination; employment—and we can go on and on.

For this new movement for environmental justice, matters of the environment are not confined to how to best manage or preserve some extra-urban "wilderness"; rather, the environment is part of a broader framework of economic, racial, and social justice. This perspective represents a significant challenge to the way mainstream environmental groups have commonly presented the environmental agenda as primarily occupied with the conservation of pristine wilderness and wildlife. The exclusion of any discussion of urban and industrial concerns in mainstream environmentalism is reflected in the histories of environmentalism, which concern themselves with the romantic champions of wild "Nature," such as the 19th century national parks advocate John Muir and Aldo Leopold, whose "land ethic" has become so revered by much of modern environmentalism, but say nothing of struggles to improve the urban and industrial environments.

By redefining environmentalism, the environmental justice movement has endeavoured to rewrite environmental discourse and restructure environmental campaigning to represent the concerns, organisational strategies, and demographics of communities who are confronted by the worst excesses of environmental degradation.

## SOME HISTORICAL LANDMARKS

Because of its community-based approach to organisation, the environmental justice movement is defined by the numerous activities of groups and individuals, each of which has its own story of its evolution to tell; nevertheless, a number of significant landmarks stand out in the development of the movement.

Although, the issues and campaigns may have been called something other than environmental justice, it is possible to discover a long history of what it encompasses. Friedrick Engels, for example, in his description of the living and working conditions of the working class in mid-19th century England, documents the "environmental racism" suffered by the Irish communities in Manchester.

Environmental justice advocates in the United States see their campaigning as a continuation and development of a long history of struggles against such "social problems" as dangerous work conditions, poor housing, and poor sanitation, although they have now become framed as "environmental problems".

The demonstrations in Warren County, North Carolina, in 1982 by a community of mostly African-Americans against a proposed PCB landfill site are held by many within the environmental justice movement to mark the birth of the national environmental justice movement. It was during these protests that the term "environmental racism" was coined.

The landmark study of what had become called environmental racism was carried out in 1987 by a civil rights organisation, the United Church of Christ Commission for Racial Justice. Entitled *Toxic Wastes and Race in the U.S.* (Public Data Access, New York), the study exposed a link between the location of hazardous waste disposal sites and the racial profile of the neighbouring communities. The study concluded that race was the most significant variable with regard to the sitting of hazardous waste industries: communities of colour had the largest number of hazardous commercial waste operations, with three out of every five black or Hispanic Americans living in communities with at least one toxic waste site. These findings were confirmed in Robert

Bullard's book, *Dumping in Dixie,* published in 1990, which documented this form of racism in the southern United States. This particular brand of environmental inequity was found by subsequent studies to operate in the location and through the detrimental impact of other environmentally hazardous industries and practices, such as the manufacture and use of pesticides, and health and safety concerns such as lead poisoning.

In 1990, the environmental justice movement sent a direct challenge to the mainstream U.S. environmental groups, known as the *Group of Ten* (which includes organisations such as the Audubon Society, the Sierra Club, and the Wilderness Society). In two open letters of protest to the Group of Ten, the environmental justice groups, the Gulf Coast Tenants Organisation, and the Southwest Organising Project, criticised the environmental movement for its racism in terms of the racial composition of the staff of the organisations making up the Group of Ten, its agenda, which failed to concern itself with poor and ethic communities despite evidence to suggest that they were the main victims of pollution, and its "clear lack of accountability" to communities of the Third World. These letters represented a protect against the institutionalisation and professionalisation of the environmental movement and its agenda, which failed to incorporate concerns of social justice.

The centrality of the link between ethnicity and those communities at greatest risk from environmentally hazardous industries and practices to the redefinition of environmentalism pursued by the environmental justice movement was registered by the first national People of Colour Environmental Leadership Summit held in Washington, DC, in October 1991. This meeting represented an opportunity for the numerous environmental justice groups and networks which had developed in the 1980s—many of which, such as the National Toxic Campaign of Boston and the Citizens Clearinghouse for Hazardous Wastes, had been making explicit links between social issues such as poverty and racial discrimination and waste disposal since the 1980s—to collectively define a new kind of environmentalism from a multiracial and social justice perspective. The summit discussed organisational questions of the movement, such as whether campaigning ought to be centralised or remain the domain of regional networks, and whether the central organisational concern of environmental racism would best be served by multiracial groups or ones whose membership exclusively represents the ethic group concerned. These debates are still current and have formed and important part of the agenda of subsequent regional follow-up meetings (the meeting in new Orleans in 1992 was attended by over 2000 people).

Importantly, delegates to the 1991 summit ratified a statement of the Principles of Environmental Justice which placed themes of sustainable development, pollution control, health and safety at work, production practices, cultural self-determintion and political empowerment, and environmental racism into a single framework, giving this new form of environmental politics a clear identity.

## COMMUNITIES OF COLOUR AND ENVIRONMENTAL HAZARDS

Benjamin Chavis of the Commission for Racial Justice has defined environmental racism as racial discrimination in environmental policy making. It is racial discrimination in the enforcement of regulations and laws. It is racial discrimination in the deliberate targeting of communities of colour for toxic waste disposal and the siting of polluting industries. It is racial discrimination in the official sanctioning of the life-threatening presence of poisons and pollutants in communities of colour. And, it is racial discrimination in the history of excluding people of colour from the mainstream environmental groups, decision making boards, commissions, and regulatory bodies.

Racism has not only shaped the social, economic, and political landscapes of countries such as the United States, it has also played a significant part in the formation of the environmental landscape of countries. Racial discrimination has meant people of colour are concentrated in communities that are segregated from white residential areas, and it is these identifiable communities of colour that are targeted for a disparate exposure to environmental risks. Environmental racism is seen by the environmental justice movement as an extension of institutional racism in housing, education, employment, and so on.

Studies conducted by environmental justice advocates, and recently by the U.S. Government's Environmental Protection Agency (in a 1992 two-volume report entitled *Environmental Equity,* EPA Office of Policy, Planning, and Evaluation, Washington, DC), conclude that communities of colour are subjected to a disproportionate share of the burden of environmental hazards. Racism has been, and continues to be, a decisive factor in environmental planning and policy making.

This disproportionate concentration of environmental hazards in communities of colour indicates that the siting of polluting waste treatment facilities, and other environmentally dangerous operations, is decided not by considerations of environmental suitability, but by an evaluation of the social power of neighbouring communities and the subsequent likelihood of effective organised opposition. Poor communities, particularly those of people of colour, with high levels of unemployment, poverty, ill health, poor housing, inadequate educational resources, and poor political representation, are considered the most likely to be able to offer the least resistance to these highly constancies and contested policy decisions. Measures by governments to tighten environmental regulations on industry have had the unconsidered effect of intensifying this burden of environmental racism by increasing the demand for waste treatment facilities.

In work, ethnicity is a factor which generally compounds problems of economic injustice. People of colour are concentrated in occupations with high risks to health. Studies, such as those edited by Bunyan Bryant and Paul Mohai (1993. *Race and the Incidence of Environmental Hazards: A Time for Discourse.* Westview, Boulder,CO), show that farm workers and their children, the vast majority of whom in the United States are people of colour, are commonly exposed to poisonous levels of pesticides. This is also true of the workers involved in the production of the pesticides and the communities, again most likely to be people of colour, who neighbour these production sites.

Not only are communities of colour disproportionately exposed to environmental risks, the effects of this exposure are unlikely to be adequately treated by a health caste system which, as studies such as Vicente Navarro's have shown, is also permeated by racial discrimination and subsequently offers a reduced and impoverished service to people of colour.

In the United States, the lands of native Americans have suffered from the attention of waste disposal industries, with reservations targeted for landfills and incinerators, although a great many of these planned developments have been frustrated by community opposition.

Less successful has been the defense of these native American communities against "economic blackmail."

Ward Churchil and Winon LaDuke have described how the Navajo and Lakota Indian Reservations have been held "economic hostages" by the uranium extraction industry while their health and environment have been decimated by the resultant radioactive waste. Poor communities generally are susceptible to "job blackmail," which offers unemployed communities the lure of jobs with the threat that should corporate demands be challenged these employment opportunities will be dissolved. In the case of the Navajo and Lakota people, the success of this intimidation has resulted in such severe radioactive pollution that the U.S. government is considering designating these, and other affected lands, National Sacrifice Areas. The implementation of this plan would effectively dispossess Indian peoples, who would be forced to leave these areas, of their homelands. A Churchil and LaDuke conclude, such a decision would effectively designate these Indian communities as "National Sacrifice Peoples."

## WOMEN AND THE ENVIRONMENT

While much of the work done in this effort to redefine environmentalism has been conducted on the issue of environmental racism, environmental injustice is not confined to discrimination on grounds of race. The framework of environmental justice examines the distribution of harmful effects of environmentally destructive practices along class and gender lines as well

A great many of the slogans of mainstream environmentalism, and their transmission by the media, rest on the assumption that our present environmental insecurity is the result of a culpable universal human endeavour to "dominate" nature. This view has important implications for the ascription of responsibility for environmental degradation. The perceived domination of nature by humanity as a whole implies a universality with informs a discourse of responsibility that blames indiscriminately. The received media expression of this perceived universal responsibility has allowed the widespread circulation of a mentality which is informed by the self-accusatory assertion, "the enemy is us." As a result a great deal of campaigning by environmental groups, sometimes with the support of governmental agencies, has been directed at the reform of individual behaviour; for example, we are given the opportunity to redeem ourselves via "environmentally friendly consumption." This "privatisation" of environmental costs has resulted in a disproportionate burden of responsibility being placed on women.

One area of environmental concern where this displacement of responsibility from the wasteful productive practices of industry to the individual is particularly obvious concerns the perceived responsibility for waste. Despite the fact that household rubbish makes up approximately only 10% of the total waste of towns, since their inception in the 1980s, waste reduction and recycling schemes have been concentrated on the household. The high-profile efforts of the state have given rise to a "waste ethic" which informs waste coordination and sorting efforts and environmental information campaigns which are relayed through consumer associations, the media, and governmental bodies. Waste education, which instructs the household to reduce waste at the point of purchase by favouring "environmentally friendly" produced and packaged goods, and at the point of disposal by sorting waste for recycling, is a strategy of displacement in which the household is not only required to absorb the consequences of inhealthy working and living conditions, but is further burdened with the responsibility for environmental recovery and sensible management of valuable natural resources and dangerous waste products.

As Irmgard Schultz notes, this state-sponsored "waste ethic" is targeted specifically at women: The goal is to create a waste consciousness among women. The method is to burden women with a guilty conscience if they do not correctly sort the rubbish...the goal of this huge information campaign is to discipline women to collect and separate rubbish, dovetailing both with the teaching of cleanliness to children and with women's overall management of the household.

This tactic of displacing responsibility for waste management onto the household, in which women ultimately bear the greatest responsibility, identifies waste as a problem of individual households whose patterns of consumption and disposal are in need of improvement and reform, and not as a problem of market-governed production principally interested in profitable sales. In this manner, the costs of environmental degradation are passed onto the household and thereby privatised. Schultz concludes that this tactic must also be seen as a "moralizing strategy" against women by maintaining that the waste problem is a consequence of women's moral behaviour, thereby pressing women to adopt responsibility for waste management.

Schultz believes that women are particularly susceptible to this environmental blackmail because of their established role as domestic caregivers. A guilty conscience about waste can be engendered in women because of an existing "gender-specific" responsibility for health. For the most part, it is women who have been allocated responsibility for the health of the household members, and it is out of a concern for the detrimental impact of waste on health that women have engaged with the demands of the "waste ethic" peddled by the state.

This moralising ethic also facilitates the co-option of

unpaid female labour into the production process, thereby aiding the accumulation of capital. To make productive and profitable use of recyclable wate products, it is necessary for the materials to be sorted prior to technological reprocessing, and this sorting of waste is performed in the private household, most often by women. This moralising strategy therefore not only displaces responsibility for waste reduction and disposal onto the private household, but also co-opts unpaid domestic labour into the production process. In this respect, the strategy is a twofold disciplinary mechanism: not only does it morally discipline women by engendering a guilty conscience about the volume and environmental impact of waste, it also represents a form of work discipline requiring the ordered sorting and disposal of recyclable products for the advantage of corporate industry.

The inequalities that women suffer in the developed world are compounded by economic vulnerability; it is poor women in the developed world, often women of colour, who are likely to suffer the consequences of a degraded and polluted environment. In the Third World gender inequality means that women suffer most as a result of the inequalities between rich and poor countries. The inequalities that women suffer in the developed world are intensified for women in the Third World.

The marginalisation of women in the public statements of environmental organisations, and the conservation strategies they evolve, is well illustrated by the inclusion of women, along with indigenous people and the young, in the classification "other groups" in Agenda 21 of the Rio Declaration on Environment and Development issued from the Rio Summit of 1992. Although, these three groups together make up the majority of the world's population, with women making up about half of that total, the male dominance of the perspective that sees women as "other" ensures their marginalisation in such prominent statements of intent as the Rio Declaration.

Yet, it is women in the Third World who are confronted by the worst excesses of unsustainable, environmentally destructive practices. Degraded agricultural land and poor and limited access to food, potable water, health care, and energy supplies—the results of the degradation of the environmental and unsustainable, environmentally destructive state and capitalist activities—have their initial and greatest impact on the lives of women and their children. This is borne out by the enormous disparities in infant and maternal rates of ill health, malnutrition, and mortality between the developed world and the Third World.

Women of the Third World have not met these injustices with passivity (despite prevailing stereotypes). Numerous women's groups exist in countries of the Third World to secure the needs of women in both rural and urban regions. These groups have not only engaged in activities which might conform to the often favoured view of women as carers, such as the defense of natural habitats and the reclamation of land, but women's groups like the Self-Employed Women's Association, founded in 1972 in Ahmedabad, have involved themselves in trade issues in pursuit of more equitable trade conditions for women. Women have also sought involvement in urban planning—as exemplified by CEFEMINA, a Costa Rican group formed in 1981 to petition for environmentally sensitive housing development—and the provision of literacy programmes, like the Chilean group MOMUPO. These groups, which have arisen to combat the environmental, social, and economic injustices that women of the Third World disproportionately suffer, are evidence that the politics of gender must be included in any credible account of an environmentally sustainable world.

Since the environmental justice movements beginnings in the antitoxic campaigns of the early 1980s, women, predominantly women of colour, have played a vital role. This membership profile rejects the fact that the majority of community groups that make up the environmental justice movement are located in working class communities in urban and industrial centres. The disproportionate representation of people of colour, women, and the poor in this movement reflects the disproportionate burden of environmental hazards these sections of the population suffer.

In these communities, the vital environmental issues are those everyday ones concerned with health and survival, and it has been suggested by some theorists that it is women's biological or social affinity with these concerns that has led to their predominance in the environmental movement generally. While some ecofeminists claim that it is a women's biological role of reproduction that informs an innate appreciation of the unity of all life and encourages them to lead the fight for a clean and healthy environment for their families, others have argued that it is women's gendered social roles as "mother" and "carer" that have determined this involvement.

However, some women campaigners in the environmental justice movement have been cautious of such ecofeminist theorising, questioning its political usefulness and strategic value; Dana Alston, for example, has stated, "Personally, I like spending my energy trying to change how we are perceived and how we more forward rather than analysing why we're there in the first place". Whatever the reason, there is no doubt that women do suffer environmental injustices as a result of their gender and that women like Lois Gibbs and Penny Newman of the Citizen's Clearing-house for Hazardous Waste have made a vital contribution to the environmental justice movement.

## LABOUR AND THE ENVIRONMENTAL MOVEMENT

The environmental movement has, for the most part, distanced itself from labour-organised opposition to the activities of industry. The loyalties and motives of labour appear to be viewed with a certain cyncism and disdain. Some environmentalists, like the deep ecologists Warwick Fox, have suggested that the working class colludes with corporate interests responsible for environmental destruction, while others, such as Ulrich Beck, have suggested the very notion of class is itself redundant. It seems that from the view point of environmentalism the working class, if it exists at all, is untrustworthy because it is corrupted by the bribes of capital and will seek to defend these interests by siding with industry against an environmentalism which may threaten them.

Similarly labour has opposed environmental protest and activities which it perceives as a threat to continued employment in industries which the environmental movement has identified as environmentally hazardous. The mutual hostility and suspicion of environmentalism and labour finds much of its motivation in the perceived opposition between the interests of labour in securing its continued means of subsistence—employment—and the interests of the environment, which may require the drastic curtailment, and even the cessation, of the activities of many Industries (the source of employment). By manipulating, the character of this discussion, capital has been able to promote the belief that regulations on industries which seek to defend the environment will lead to losses in employment, thereby contriving the myth of "jobs versus environment," which translates as "labour versus environmentalists" and creates deep divisions between labour and the environmental movement.

Labour's struggle for improved work and living

"labour versus environmentalists" and creates deep divisions between labour and the environmental movement.

Labour's struggle for improved work and living conditions has been a continuous element of class conflict. Demands for a healthy environment, both inside and outside the factory, and challenges to the operating procedures of industrial capital have been a central component of the historical opposition of labour. Capital in turn has sought to resist the internalisation of the costs of its destructive practices. These costs are presently borne by society (most severely by the working class and racial minorities, and, with the international mobility of industrial capital, by the poor of the third World) in the form of pollution, occupational disease, and widespread public ill health. The economic burden of these consequences of capitalist production is met by social institutions, such as health services, where these exist, but otherwise directly by the individual and his or her family and friends.

The unwillingness of industry to take account of the costs of its operations to the environment, communities, and workers is not solely driven by economic considerations; regulations designed to protect the environment, worker, and public health are resisted because they increase worker and public involvement in production and investment decisions or private industry, and represent an interference with the jealously guarded "exploitive rights" of capital. While capital is successful in fending off these challenges to its authority, businesses collect, as Richard Kazis and Richard Grossman describe it, "a subsidy coined in blood, broken bones, broken health, physical pain, mental anguish and the ultimate trauma of death" (1991).

The poor suffer disproportionate exposure to environmental hazards at work (when they are able to get work), whether it be in the production of toxic substances or in their use, and in their communities, which are most likely to be the location of polluting industries. For the working class, health and safety at work, involvement in production decisions, and community conditions are environmental issues. The framework of environmental justice recognises the essential relationship between these concerns of the working class and environmental problems.

Efforts to include labour into the environmental justice movement have centred around the issue of job blackmail; employers are able to intimidate workers by threatening to close factories and move to alternative sites elsewhere in the country or abroad, leaving unemployment in their wake. This intimidation is also levelled at communities who are told to conform to corporate requirements if they are to retain existing employment or attract new opportunities.

With increased pressure on industry to clean up its practices, this ultimatum is ever more frequently made to counter public demands for environmental and occupational health and safety protection. By a clever public relations exercise, which focuses on the employment benefits which its present, or proposed, investment promises, business is able to present its own private interests as synonymous with the public's interest: the pursuit of profit dons the benevolent mask of a public service—the creation of employment opportunities. This lure of jobs is reinforced by the threat of unemployment devised to intimidate workers into accepting that their only alternative, if they are to avoid the economic hardship of unemployment, is to support the proposals of employers in spite of their detrimental and exploitative character. This job blackmail represents economic bondage and is used by industry to define the terms of labour's good behaviour. Aware of the brutal consequences of its market-governed performance, in both human and environmental terms, business seeks after workers have sacrificed environmental and health and safety regulations for continued employment, poor health resulting from occupational diseases and possible untimely unemployment or death resulting from occupational accidents become a fearful reality. Insurers, quick to spy the supply-side advantages of worker's concern in the face of such threats, have eagerly offered the "solution" of health and life insurance. By tying such schemes to employment, capital is able to further subjugate labour and police its good behaviour. This move seems to go hand-in-hand with government underfunding and privatisation of public health care provision.

By manipulating the public's perception of possible alternatives, corporations and their supporters are able to imposed the terms of the debate. Speaking of the clash of interests between the northwest American timber industry and environmentalists concerned to preserve the forest habitat of the endagered spotted owl, then President George Bush stated, "We want to save the little furry-feathery guy and all that but I don't want to see 40,000 loggers thrown out of work". Such "definitional sleights-of-hand" allow governmental officials and business to avoid basic considerations of what it is that is being produced and for what purpose, and how, if at all, it should be produced. The contrivance of this false conflict between jobs and the environment provides capital with the opportunity to continue with "business as usual."

In an effort to create a link between employment and the environment, environmental justice groups, such as the National Toxics Campaign, proposed the establishment of a "superfund for workers." This concept recognises the importance of job loss to both labour and environmental groups by calling for financial and training assistance for workers who become unemployed as a result of the termination of environmentally hazardous industries.

Similarly, environmental justice calls for the restructuring of industry so as to reduce environmentally harmful production and requires the support of labour initiatives, such as that of the Lucas Aerospace Campaign in the 1970s, which sought the opportunity for workers to turn their skills and equipment to the production of environmentally benign, socially useful products.

The framework of environmental justice recognises that the degradation of the environment is sourced in the procedures of production. It is the organisation or production in society, with its relations of exploitation, that is the object of the environmental justice critique. Current production is organised for the realisation of profit and privilege, and pays little heed to the consequences for the health of humanity or the rest of nature. The impact of environmentally hazardous products on wild nature is preceded by their production and their detrimental effects on the health and quality of the lives of workers and their communities during this production. Environmental justice appreciates, therefore, that the terrain of production and the support of labour are fundamental to environmental campaigning.

Thus, the link between the interests of labour and the interests of the environment is forged by the realisation that environmental issues are questions about the social organisation of production; environmental justice dictates that it is not enough to ask, "what is produced?", but also "How do we produce?", and "Who is in control of this production?" Concern about what kind of environment we want to live in presupposes concern about how we labour. The environmental justice movement argues that at present the interests of profit and privilege determine the answers to these questions; it campaigns for justice to determine them.

## THE THIRD WORLD AND THE ENVIRONMENT

The struggle for environmental justice is an international one reflecting the global character of environmental problems. The Third World has become a dumping ground not only for the wastes of developed countries, but also for products considered too hazardous for the domestic market, including pesticides and medicines. The poorer countries of the world have also received the attention of corporations seeking cheap labour and a way of avoiding costly environmental and health and safety regulations on their operations; the great loss of life at Bhopal is a testament to the enormous dangers Third World peoples face as a result of these cost-cutting exercises.

As restrictive regulations on industrial practices and the disposal of waste have been applied in the industrialised countries, companies have sought out chapter alternative sites in the Third World, where regulations are nonexistent or ignored, the cost of wâste disposal in the United States, for example, may be over 20 times more than that in a country of in Africa. As the regulation of toxic waste disposal has tightened and the demand for legitimate domestic sites has outstripped provision, companies have found a cheaper and largely unrestrained alternative abroad in countries of the Third World.

Some of the international trade in waste is carried out legally with Third World waste importation companies and governments eager to earn foreign currency. The Third World Network, who monitors this trade, identities two "waves" of such waste disposal contracts:

The first, in 1980, involved the U.S. and coincided with the tightening up of laws concerning toxic waste there. The second wave in 1987 and 1988, involving European countries, was probably linked to the 1986 EEC directive and decisions arrived at by the Organisation of European Cooperation and Development (OECD) on the transportation of toxic waste.

There is also complementary, and widespread, illegal trade in waste often involving the unscrupulous complicity of governments of the developed and underdeveloped worlds.

This "toxic imperialism" represents the globalisation of the environmentally racist practice of targeting domestic communities of colour for the siting of environmentally hazardous facilities, as uncovered in the United States in such studies as the Commission for Racial Justice and those of Robert Bullard. The institutional nature of this racism was recently made explicit in a leaked memo from Lawrence Summers, chief economist of the World Bank. Part of this memo was published in *The Economist* in February 1992 under the title, "Let Them Eat Pollution." The memo presents an "economic" argument for an increased trade in toxic waste between the developed world and the Third World based on such reasoning as the differential loss of earnings resulting from health-impairing pollution between these two populations, favouring an increased level of pollution in countries with the lowest wages, and the "inefficiently low" levels of pollution in African countries compared with those of the developed world.

## CONSERVATION AND JUSTICE

There is increasing concern within the environmental justice movement that people of the Third World are suffering injustices resulting from the conservation strategies promoted by mainstream environmental groups and governments of the developed world. A good example of this are the "debt-for-nature" exchanges initiated by the World Wildlife Fund (now the Worldwide Fund for Nature), and sponsored by the United States government which enable transnational forces—economic and cultural—to penetrate the political economies of Third World countries and secure control over the conservation programmes of these countries, often at the expense, and to the detriment, of indigenous peoples. Since, the first debt-for-nature exchange in 1987, when protection for 3.7 million acres of Bolivian tropical forest was negotiated in return for money raised by Conservation Internal (from public and private financial institutions, governments, and the United Nations) to buy $650,000 of Bolivian debt, such swaps have been arranged, or considered, in a large number of countries, including Brazil, Argentina, Ecuador, Venezuela, Gautemala, Zambia, the Dominican Republic, and Jamaica.

While it does appear that this initiative has extended some protection to an extremely limited and select portion of the biosphere, what is clear is that these new relations between conservationists and the international financial community have seriously undermined the struggles of indigenous people for self-determination and control of their homelands. This concern was addressed by the Coordinating Body for Indigenous People's Organisations of the Amazon Basin (COICA), which represents 1.2 million Indian people in Colombia, Peru, Bolivia, Ecuador, and Brazil, in an open letter to the international conservation community of 1990:

We are concerned about debt-for-nature swaps that put your organisations in a position of negotiating with our governments the future of our homelands. We know of specific examples of such swaps which have shown brazen disregard for the rights of the indigenous inhabitants...

We want to make it clear that we never delegated any power of representation to the environmentalist community nor to any individual organisation within the community.

We propose joining hands with those members of the

world-wide environmentalist community who recognise our historical role as caretakers of the Amazon Basin, support our efforts to reclaim and defend our traditional territories, and accept our organisations as legitimate and equal partners. It seems that the economic vulnerability of these countries, on which this programme trades, is likely to lead to financial expediency suppressing matters of social justice.

## ENVIRONMENTALISM AND DEVELOPMENT

Environmentalism as pursued by the domestic governments of countries of the Third World is supported to an ever increasing extent by the development assistance of foreign agencies. Such foreign-sponsored environmentalism serves to reinforce dependency relations between poor countries of the Third World and the wealthy developed nations. Not only do environmental conservation and rehabilitation programmes cost enormous sums of money, which a debt-burdened Third World cannot afford, but with the increasing "industrialisation" of environmentalism, and its monopoly of "expert" technologies of assessment, reconstruction, and management, industrial nations are able to regulate the conservation of the natural resources of the Third World and manage their exploitation. Such a relation of financial and technological dependency means that the economies of the developed countries are able to determine the agenda of the environmental policies of Third World countries so as to control global environmental resources and secure continued, and often improved, access to them. In this manner, environmentalism represents another avenue for the insertion of political economies of the Third World into the global corporate system. By reinforcing relations of dependency, environmentalism becomes another example of "dependent development" which enables developed nations to discipline the political economies of countries of the Third World.

## CONSERVATION PROGRAMMES AND INEQUALITY

The discourse of environmentalism has also been pressed into service by the domestic elites of countries of the Third World to legitimise the reproduction of social inequalities. State-led environmentalism frequently identifies the poor as responsible for environmental degradation: condemnation by vocal elites is levelled against the sexual practices of the poor in their failure to reduce family size and the subsistence practices of the poor, such as the clearance of forest for cultivation and collection of firewood for cooking. In this fashion attention is diverted from the role that corporate interests play in the destruction of the environments of countries of the Third World by their intensive logging, fishing, and agricultural industries, while collaborative domestic elites further enhance their economic wealth.

It is the poor, who suffer the brunt of environmental degradation, who are identified by state environmentalism as to blame. Subsequent elite formulated and foreign-funded responses to the environmental crisis comprehend and objectify the poor only as "passive participants." for example, as paid labourers in reforestation programmes. The poor are dispossessed and proletarianised as they are forced out from degraded environments which are no longer sufficiently productive into paid labour, often recruited by elite-led, foreign-funded development programmes.

These elite-driven environmental protection programmes act as a disciplining mechanism against the poor. These measures may often involve incorporating poor farmers into the "Green Revolution," with the result that food production becomes increasingly depended on the technologies of chemical pest control, chemical fertilizers, agricultural machinery, and genetically engineered seed varieties, control and distribution of which is monopolised by corporations of the developed world. Moral discipline of the rural poor, directed at their sexual practices, is thereby closely accompanied by the disciplining of their agricultural practices and the creation of a dependency on the technologies of the developed world.

## WILDERNESS DESIGNATION AND "ECOTOURISM"

Conservation initiatives such as the debt-for-nature swaps are also pursued with the intention of defending the "wilderness" as a leisure opportunity for the wealthier people of the developed world. With a number of tour operators taking an active involvement in conservation programmes, the idea is to develop a "low-impact" tourism, or "ecotourism," in countries of the Third World. Ian Munt has disclosed the implicit racism of the primitivistic nostalgia surrounding the romantisation of wilderness and native people which informs the rhetoric of this tourism, with its frequent employment of colonial imagery and recreation of forms of colonial travel, such as safaris, in a new "ecofriendly" guise as "ecosafaris".

The designation of wilderness preserves in the Third World is resulting in severe injustices and deprivation. Ashish Kothari, the founding member of the Indian environmental group Kalpvriksh, says of the establishment of the national parks in India. The approach was essentially one of saving wildlife *from* the people, rather than *with* them. In other words, local communities in wildlife-rich areas were seen to be enemies to the conservation cause, and were either physically displaced or denied access to these areas. While in the short run this approach has led to habitat and species protection in several parts of India, it has also had two detrimental impacts: first, severe human rights abuses as people's basic survival resources have been denied them; second, the alienation of local people from the very areas which they could have been instrumental in conserving. The approach is therefore not only anti-human, it is also short-sighted from the wildlife conservation point of view. Increasingly, as sharp backlash, communities are demanding the denotification of national parks and sanctuaries in order to regain their rights over resources.

The Third World poor, dependent on the lands subject to wilderness designation for their means of survival, cannot afford the luxury of a "museumised nature." Conservation efforts in the Third World which are conducted to the exclusion of the welfare and interests of indigenous communities are unjust insofar as they have led to the dispossession of people of their lands, and subsequently their means of survival. Environmental justice demands equitable access to, and distribution of, the resources of nature, in a manner which is sustainable for present and future generations.

## INTERGENERATIONAL AND INTERSPECIES JUSTICE

The demands of justice, at their very least, are met when the basic needs of a person which enable one to live a life that is worthwhile and devoid of such harms as hunger, malnutrition, ill-health, and a polluted environment are satisfied (where this is possible). It is unreasonable, though some have suggested it, to assume that relations of justice and injustice are bounded by the domestic concerns of any particular country or culture, or the limited obligations defined by relations of reciprocal advantage, or some special relationship, such as those of the family. If one persons basic needs are to count, then the demands of moral equality require that those of anyone else count equally, regardless of their geographical location, or relationship or lack of it, to any other person.

The pertinent question in considerations of just treatment is, "do they have needs and interests of moral significance which may be aided or frustrated by our actions?" If this is the requirement for relations of justice to exist, then, as is recognised in "Principles of Environmental Justice", their sphere of application is not confined to the present population of human beings.

It is widely accepted that our present actions may have serious repercussions for future generations of human beings. It does not take much imagination to recognise that the legacy of a polluted, exhausted, inhealthy environment will significantly frustrate the satisfaction of the interests of future generations and their ability to meet basic needs, such as clean air and water, sufficient energy supplies, and a healthy environment in which to live, and the fact that the identities of future people are currently unknown does not affect this recognition. Justice, therefore, requires that we consider the needs and interests of future generations when evaluating the acceptability of present actions, like those that have a detrimental environmental impact.

The realm of justice does not end with the treatment of humans, although there is considerable debate about what in the natural world is worthy of moral consideration, whether it be sentient animals or all animals, whether plant life ought to be included in the moral community, or whether it is collectives such as species or ecosystems which ought to be the objects of our moral concern. While these debates in environmental ethics and animal welfare ethics are ongoing, it does seem clear that nonhuman species do have basic needs which certain of our actions (particularly those of an environmentally destructive character) can frustrate and to which we therefore own concerns of justice.

Environmental justice not only mandates that we ensure

a healthy environment for present and future generations of human beings, but also for present and future generations of nonhuman life.

## PRINCIPLES OF ENVIRONMENTAL JUSTICE

### Preamble

We the people of colour, gathered together at this multinational People of Colour Environmental Leadership Summit, to begin to build a national and international movement of all people of colour to fight the destruction and taking of our lands and communities, do hereby re-establish our spiritual interdependence to the sacredness of our Mother Earth; to respect and celebrate each of our cultures languages and beliefs about the natural world and our roles in healing ourselves; to insure environmental justice; to promote economic alternatives which would contribute to the development of environmentally safe livelihoods; and, to secure our political, economic and cultural liberation that has been denied for over 500 years of colonisation and oppression, resulting in the poisoning of our communities and land and the genocide of our people, do affirm and adopt these Principles of Environmental Justice.

1. *Environmental justice* affirms the sacredness of Mother Earth, ecological unity and the interdependence of all species, and the right to be free from ecological destruction.

2. *Environmental justice* demands that public policy be based on mutual respect and justice for all people, free from any form of discrimination or bias.

3. *Environmental justice* mandates the right to ethical, balanced and responsible uses of land and renewable resources in the interest of a sustainable planet for humans and other living things.

4. *Environmental justice* calls for universal protection from nuclear testing and extraction, production and disposal of toxic/hazardous wastes and poisons that threaten the fundamental right to clean air, land, water, and food.

5. *Environmental justice* affirms the fundamental right to political, economic cultural and environmental self-determination of all people.

6. *Environmental justice* demands the cessation of the production of all toxins, hazardous wastes, and radioactive materials, and that all past and current producers be held strictly accountable to the people for detoxification and the containment at the point of production.

7. *Environmental justice* demands the right to participate as equal partners at every level of decision-making including needs assessment, planning, implementation, enforcement and evaluation.

8. *Environmental justice* affirms the right of all workers to a safe and healthy work environment, without being forced to choose an unsafe livelihood and unemployment. It also affirms the right of those who work at home to be free from environmental hazards.

9. *Environmental justice* protects the right of victims of environmental injustice to receive full compensation and reparations for damages as well as quality health care.

10. *Environmental justice* considers governmental acts of environmental injustice a violation of international law, the Universal Declaration On Human Rights, and the United Nations Convention on Genocide.

11. *Environmental justice* must recognise a special legal and natural relationship of Native Peoples to the U.S. government through treaties, agreements, compacts, and covenants affirming sovereignty and self-determination.

12. *Environmental justice* affirms the need for urban and rural ecological policies to clean up and rebuild our cities and rural areas in balance with nature, honouring the cultural integrity of all our communities, providing fair access for all to the full range of resources.

13. *Environmental justice* calls for the strict enforcement of principles of informed consent, and a half to the testing of experimental reproductive and medical procedures and vaccinations on people of colour.

14. *Environmental justice* opposes the destructive operations of multi-national corporations.

15. *Environmental justice* opposes military occupation, repression and exploitation of lands, people and cultures, and other life forms.

16. *Environmental justice* calls for the education of present and future generations which emphasises social and environmental issues, based on our experience and an appreciation of our diverse cultural perspectives.

17. *Environmental justice* requires that we, as individuals, make personal and consumer choices to consume as little of Mother Earth's resources and to produce as little waste as possible; and make the conscious decision to challenge and reprioritise our lifestyles to insure the health of the natural world for present and future generations.

Adopted October 7, 1991, National People of Colour Environmental Leadership Summit, Washington, DC.

## REFERENCES

Alston, D. (Ed.), "We Speak for Ourselves: Social Justice, Race, and the Environment." The Panos Institute, Washington, DC.

Beder, S. (1993), *The Nature of Sustainable Development,* Newham, Australia: Scribe Publications.

Bryant, B., and Mohai, P. (Eds.) (1993), "Race and the Incidence of Environmental Hazards: A Time for Discourse." Westview, Boulder, CO.

Bullard, R. (1990), "Dumping in Dixie: Race, Class and Environmental quality," Westview, Boulder, CO.

Bullard, R. (Ed). (1993), "Confronting Environmental Racism: Voices from the Grassroots," South End, Boston.

Churchill, W., and LaDuke, W. (1985), Radioactive Colonisation and the Native American, *Socialist Review*, 81, 95—119.

Cropper, M.L., & Oates, W.E. (1992), Environmental economics: A Survey, *Journal of Economic Literature*, 30, 675-740.

Di Chiro, G. (1992), Defining Environmental Justice: Women's Vioces and Grassroots Politics. *Sociatist Rev.* 44(8), 10-20.

Fortune, L. (1994), Environmental Management in Packaging: Responsponding to the Dream, In D. Shillito (Ed.), *Implementing Environmental Management: Policy and Techniques*, Rugby, UK, Institution of Chemical Engineers.

Freeman, A.M. III. (1993), *The Measurement of Environmental and Resource Values*, Washington DC: Resources for the Future.

Gotlieb, R. (1993), "Forcing the Spring: The Transformation of the Environmental Movement", Island, Washington, DC.

Gray, R., Owen, D., & Adams, C. (1996), *Accounting and Accountability: Changes and Challenges in Corporate Social and Environmental Reporting*, London and New York: Prentice Hall Europe

Hausman, J.A. (Ed ), (1993), *Contingent Valuation: A Critical Assessment*, Amsterdam: North Halland Press.

Horichter, R. (Ed.) (1993). "Toxic Struggles: The Theory and Practice of Environmental Justice," New Society Publications, Philadelphia.

House of Representatives Satanding Committee on Environment and conservation (HRSCEC), (1987), "*Fiscal measures and the achievement of Environmental Objectives*" Canberra, Australis, AGPS.

Kahneman, D. & Varey, C. (1991), Notes on the Psychology of Utility. In J. Elster and J. Roemer (Eds.), *Interpersonal Comparisons of Well-Being,* pp. 127-163, New York: Cambridge Unviersity Press.

Kazis, R., and Grossman, R.L. (1991), "Fear at Work: Job Blackmail, Labour and the Environment," 2nd ed. New Society Publications, Philladelphia.

March, J.G. (1994), *A Primer in Decision Making,* New Yrok, The Free Press.

Munt, I. (1994), Eco-tourism or Ego-tourism? *Race Class* 36(1), 49-60.

Navarro, V. (1991), Class and race: Life and Death Situations, *Monthly Rev.* 43(4), 1-13.

Organisation for Economic Co-operation and Development, (1994a), *OECD Environmental Performance Reviews,* United Kingdom and Paris.

Organisation for Economic Co-operation and Development, (1994b), *The Distributive Effects of Economic Instruments for Environmental Policy,* Paris.

Perrings, et al. (1995), Economic Values of Biodiversity, pp. 823-914, In Heywood, V.H. (Ed.), *Global Biodiversity Assessment,* pp. 823-914, Cambride, Cambridge University Press.

Sagoff, M. (1994), Should Preferences Count? *Land Economics,* 70(2), 127-144.

Shiva, V. (1989), "Staying Alive: Women, Ecology and Development." Zed Books, London.

Stone, C.D. (1993), *The Gnat is Older than Man: Global Environment and Human Agenda,* Princeton, NJ: Princeton University Press.

Third World Network (1989), Toxic Waste Dumping in the Third World, *Race Class* 30(3), 47-56.

Tietenberg, T.H. (1994), *Economics and Environmental Policy,* Brookfield, VT: E. Elgar.

U.S. Environmental Protection Agency, (1991, March), *Economic Incentives: Options for Environmental Protection,* Report of Economic Incentives Task Force, Ch. I. Washington, DC: EPA Doc. # 21P-2201.

Willums, J. O., & Goluke, U. (1992), *From Ideas to Action: Business and Sustainable Development.* Oslo, Norway: International Environmental Bureau of the International Chamber of Commerce.

World Commission on Environment and Development, (1987), *Our Common Future* (The Bruntland Report), Oxford, UK:Oxford University Press.

# 12

# ASSESSMENT OF ENVIRONMENTAL IMPACT

It is one of the main tools used in current environmental management to produce sustainable development. Sustainable development has most often been defined in terms of economic development and biological reproduction. In its narrow sense, it refers to the balance of economy and environment, but new formulations also emphasise free intellectual activity and development of cultural diversity. The authors of this entry perceive that the wider interpretation of sustainable development includes a moral obligation to transmit a viable living environment to future generations. This moral obligation can be treated as part of the Social Impact Assessment (SIA) related to EIA processes. This involves basically the problem of promoting and integrating a layperson's perspective into the experts' activity in Environmental Impact Assessment (EIA). The policy of small action by the common citizens and other activities referring to the new norms of ecosocial morality should be considered seriously as social resources for sustainable development and for EIA processes.

Environmental impact assessment is one of the main tools used in current environmental management to produce sustainable development and to prevent the accumulation of environmental risks. "EIA is a *process,* a systemic process that examines the environmental consequences of development actions, in advance". Taking a technical stance, the EIA process can be analysed by its steps (e.g., identification of key impacts and prediction of impacts on natural environment) in terms of its efficiency in physical and biological risk minimising. Yet a broader view of EIA refers to the comprehensive idea of sustainable development and even to the capacities of the civil society to act as a social agent in building a sustainable heritage for future generations.

This entry does not describe in detail the process and potentialities the EIA has in technical risk minimising; neither do we argue that the current EIA processes are "unethical" or valueless. Our aim is merely to propose that the concept of ecosocial morality be considered as an integrative ethical element referring to the social aspect of EIA and to the concept of sustainable development, which is the ultimate aim of EIA processes.

Sustainable development has most often been defined in terms of economic development and biological reproduction. In its narrow sense, it refers to the balance of economy and environment. The concept of sustainable development was used already at the time of the Cocoyoc declaration on environment and development in the early 1970s. They concept gained wide popularity after the Brundtland Commission's 1987 report *Our Common Future* (Oxford: Oxford Univ. Press), in which sustainable development was defined as a tool for economic growth and equity among the world's nations. The United Nations Conference on Environment and Development (UNCED) in Rio de Janeiro in 1992, in turn, gave a new emphasis to this concept. *Agenda 21*, which was formulated during the conference, comprised three operational dimensions; an ecological dimension, a social dimension, and a cultural dimension. Hence, in addition to the Brundtland Commission's definitions of ecological and socially just economic development, the new formulation of the concept also emphasised free intellectual activity and development of cultural diversity.

As the authors of this entry perceive it, the wider interpretation of sustainable development refers to the totality of environmental and social policies and their practical implications. The basic argument we propose is that there is a moral obligation inherent to the concept of sustainable development: we should do at least as well for our successor generations as our predecessors did for us. We refer to the responsibility to transmit the common patrimony to an unspecified number of generation as understood, e.g., by Francois Ost.

Discussion on environmental ethics has previously led to a dramatic division of anthropocentric and biocentric interpretations. According to the anthropocentric approach, humans are the sole moral and valuable entity, and therefore nature is considered valid only when it is related to our goals and aims. Based on this premise, all values connected to nature are derived from human beings. Biocentric environmental ethics, in turn, argues that nature has internal values in itself which make nature equal to humans. Furthermore, our attitudes toward nature should create a more respectful behaviour in relation to nature.

Humans, as individuals and communities, assume values and bahaviour in everyday life. Ways of life can be understood along anthropocentric and biocentric lines of reasoning. An anthropocentric position is represented by ways of life pursuing the well-being and happiness of human kind as ultimate aims. The biocentric position is represented by ways of life dedicating intrinsic value to nature and to populations of various species, as exemplified in the discourse of Deep Ecology.

In a study of self-regulation in modern society, there are at least two levels at which we can meaningfully use the conceptual tools of the social sciences to examine the relationship of changing ways of life to the environment. (The concept of the way of life is considered here as a process of every day life which is based upon its essential social conditions and people's perspectives. People's perspectives refer here to the subjective goals, aspirations, and visions attached to their activities. On the one hand, there is the question of what kind of environment individual people can and will produced for themselves through everyday life choices. On the other hand, it is important to understand how the individual, as an integral part of nature, adapts and adjusts to changes in nature, and what sort of impacts and interactions are caused by human activity on and in different eco-systems.

The first level of recognition refers to the human—environment relationship, which starts with considering how people can interact with their environment without causing serious harm to the resolves, to others, or to nature. Promoting sustainable development implies recognition of the ecosocial point of view in all development efforts. Robert Tessier refers to a shift in motivation that is expected to favour the community over individual survival or interest. In analysing the concept of sustainable development Tessier even refers to collective instead of individual salvation,because he finds structural similarity in the rational action inspired by "sustainable development" and Max Weber's Protestant ethic.

As regards the second level, the key thing is to look at the foundations of the ecological self-organisation of way of life. Edgar Morin and Anne-Brigitte Kern have said that one relevant measure for a human society is whether it operates like a living machine rather than a built machine. A living machine is capable of self-corrective action, in spite of its functional deficiencies, whereas a built machine needs external intervention in an event of any malfunction. The same principle of ecological self-organisation could be thought to apply to people's everyday way of life.

The discussion on environmental ethics has only recently been connected seriously to the discussion on Western ways of life and consumption. In the 1980s, the concept of sustainability referred more to the environmental stress and catastrophes resulting from incresed population and industrialisation in the developing world or from such accidents as Chernobyl. The present discussion focuses more on the effects of daily life of an ordinary Western citizen and refers to the aggregated Environmental and social impacts of the consumption culture.

As far as large-scale environmental risks are concerned, a major problem lies in the difficulty of predicting the ecological impacts of human activity. The environmental sociologist faces the additional difficulty of predicting human activity in modern society. We still know fairly little about how and why people more about in their daily labyrinths: we know much more about the restrictions that limit the scope of our choices and about what members of modern society can do and what they leave undone. For many people, it is difficult not to take the car out of the garage every morning, even though they are well aware that the catalyser does not completely eliminate the environmental impacts of driving. On a very general level, then, we can conclude that ways of life are structured by modern economic and social systems that tend to mobilise interests and rational action that are not fully conscious of environmental risks inherent to them. Yet each citizen may, as a moral being, take a critical position on the risks induced by his or her choices, or on the reproductive policies of one's community as regards the environmental and social impact of daily decisions.

Ervin Laszlo's caricature of modern Man describes him as an accumulation of false beliefs about his chances of controlling nature. In his critique, Laszlo says that the idea of us being able to externalise the environment derives from the belief that nature is an inexhaustible reserve of raw materials. Laszlo is also sharply critical of urbanisation and the urban way of life, drawing attention to the pompous aspects of a consumer society's way of life,which are products *not* of the independent logic of economy, but of human greed, short-sightedness, and oversised desires.

The French philosopher Michel Serres goes even further when describing the unbalanced relation between humans and nature. Serres argues that our relation to the environment is parasitical; we use environmental resources, reaping all the rewards without giving any returns. Modern social structures formulate their own justification: the fundamental misbelief that human rights be accorded "every human being." According to Serres this is the very essence of the parasitical relationship—there exists no justification which includes the whole global system as a value in the final assessment.

Serres argues that in order to establish a new balance in human societies' relation to their environment, the "social contract" should be replaced by the "nature contract," in which the environment is an equal counterpart to humans. In *Le Contract Naturel,* Serres presents the idea of this symbiotic relationship, which is based on the ecosocial morality of the unique human being.

The policy making processes of modern societies reproduce elements of the social contract that refer back to the constitutional history of modern nation states. In those contracts nature is not included as a inherent entity of vital interest or of intrinsic value. Traditionally, in modern societies social norms refer to the social contract, where nature remains an external entity to be conquered and exploited. Form the sociological point of view, Serres' idea of a symbiotic relationship between man and nature, the nature contract, implies development of new norms by the ecosocial dynamics of the present civil society. This, in turn, addresses the final moral responsibility of the social community to develop the natural contract. This moral call is universal in the sense that each community has a common heritage, including the environment, to be maintained or restored not only for the benefit to us, but also for that to the future generation.

For the common person, environmental changes appear as global and comprehensive entities that are often difficult to reduce into a chain of causes and effects. If a community is to aim at a critical review of the effects of citizens' everyday life at the community level, it has to evaluate its common heritage of ecological and social ways of life. In addition, the community should be able to project impacts of any major development on the living environment, including the impact of the human—nature relation in the particular community. According to Mary Douglas and Aron Wildavsky, committees actually have a strong tendency to define impacts of change as risks that are morally loaded. However, in self protective communities risks may become "village secrets" if no procedural evaluative practice is created.

The process of perceiving risk and of taking resonsibility of the common heritage by the community is specific to particular cultural forms. Anthony Giddens refers to the culture of modern societies as post-traditional (1994). In post-traditional cultures, risks are not defined "domestically," but in the situation where impulses from the outside have a generative function. Giddens compares today's experts to traditional guardians who formulated local wisdom. But expertise, from the angle of the community, is a much more situational role that has to be tested not only in terms of knowledge but also in respect to the moral values it represents. In the final analysis, then, what is important is the trust between expert and lay person which may or may not be confirmed in any process of "environmental impact assessment." Thus, the assessment should include as its constituent element a building up of a common ecosocial morality that generates new patterns of reflexive modernity in societies which seem to have lost much of their traditional capacity for guarding the common heritage.

Even in a post-traditional society, there is a sense of solidarity with future generations. This aspect has been underlined by the discussion of sustainable development, even if it is often neglected in experts' assessments finding short cuts in practical technologies to deal with environmental problems and "development." Particularly in famlies, children represent the potential victims of environmental risks. Their children's fate make parents reflect on the threat of environmental risks and consider how long "we (our descendants) can be safe."

Many manifestations of environmental risks (such as depletion of ozone layers, the accelerated greenhouse effect, and scarcity of potable water) have led to the conclusion that risks of human impact on nature are not only local but also global by character. Correspondingly, there has been strong agreement that principles of environmental policy should be agreed upon at the level of international politics or even defined by the "global community.". The critical point for globalisation of environmental risks and the call for sustainability and the nature contract is whether ecosocial solidarity may be raised high enough to result in a new moral comprehension at the world level. According to this comprehension, factors which were considered as "development" in the past may not be considered as such any longer.

In relation to poverty, the controversy between economy and ecology is of vital interest. Particularly, the "developing countries" have had difficulties in finding sustainable options that serve their own interests and that meet with critical Western projections (as manifested, e.g., by "green" movements). Nevertheless, there has appeared a new kind of a solidarity concerning "distant others," not only the members of future generations, but also individuals who live far away. This kind of solidarity has increased simulataneously with the globalisation of environmental risks, but has also facilitated the globalisation of culture (mass media, exported cultural products, etc.). Nowadays many people are aware that "their day-to-day actions are globally consequential". Social morality thus acquires a new dimension of global solidarity that may restructure the way of life of the individual or even of the community.

Citizens' response to the challenge of changing their way of life through their own choice is in most cases a policy of small actions. Apart from those who choose a radical alternative of asceticism, citizens show their solidarity toward distant others through small actions that do not reverse their own way of life (such as recycling, rejecting harmful products and packaging, and saving electricity and gas). Yet, no matter how minute the acts may appear even to the actors themselves, they lead to the formulation of a new ecosocial norm in civil society. The formulation of ecosocial norms in the daily patterns of action indicates that the personal assessment of environmental changes my lead to new ecological adaptation—or defense mechanisms—from the bottom up and not only by top-down experts' strategies. For example, in Finland citizens commonly believe that they may affect their own ways of life and, to a certain extent, even introduce to others ecosocial morality. New ecosocial norms, are, therefore, increasingly a part of daily discussions in families, workplaces, and schools, as well as of other everyday interactions.

The application of the personal nature contract and the policy of small actions creates a new eco-social moral community which emphasises frugality as a civil virtue The interpretations of frugality may vary from conserving a household's electricity to choosing second-hand shopping. The main idea is that everyone is able to make some of these choice and become a member of the new ecosocial community. Some acts are practised very consciously as symbolic manifestations of an ecosocial orientation. In Finland, two of these widely adopted representations are giving up the use of plastic bags (used by the majority) in daily shopping or using the bicycle when going to work. Based on these symbolic actions, the members of the ecosocial community can recognise each other in the middle of daily routines.

The policy of small actions reflects citizens' orientation to a new possible future in which society has chosen a more ecological direction of development. It is considered more valuable as an act of human solidarity to "do even something" in the face of an unsecure future than just to wait and see what will happen. Even if there is no radical change of the frames or patterns of ways of life, ecosocial norms gradually generate new environmental awareness. Instead of the externalise threat of nature expressed by the agonising relation, "nature versus me," as many come to realise the globalised risks of living as we have, people will start to think in terms of "I and nature" to prepare more or less consciously a reformed basis to pursue community-level sustainability. Consequently, the policy of small actions and acts of frugality can be seen as elements of a more general ecosocial self-regulation of everyday practices. This social development seems to be crucial for the transition of awareness and action that will bring the issue of environmental reproduction and impact assessment to the community level.

There is, however, a basic problem of legitimacy and of political efficiency inherent to citizens' policy of small actions. The everyday norms of ecosocial morality and their strong attachment to the daily policy of small actions keep citizens' world of living *(Lebenswelt)* at a considerable distance from the society's environmental politics, which remains to a great extent an experts' "playground". It can also be argued that the member of ecosocial moral communities do not necessarily identify themselves with environmental movements which have become familiar through the media. Frugality and prudence as commonly appreciated virtues of civil society lead to a situation where, for example, Greenpeace's acts—if they are considered to be legal at all—gain only passive support by memebers of the ecosocial community. Legality is considered not only in social movements' activities, but is also in personal norms. environmental actions ought to meet the demands of legality. Consequently, ecosocial moral norms do not, in fact, replace traditional social norms referring to the social contract. Rather, the norms deduced from the nature contract, as perceived by laypeople of the post-traditional society, should be added to the "contract basis" that sets norms and values in social policy and in everyday life.

The formulation of the ecosocial moral community follows a certain reasoning in which the principles for personal action are formulated through an individual implementation of the "nature contract." Personal frugality and a policy of small actions may be based on a generalised moral obligation without any further development of the rational strategy that is usually called for in societal environmental policy and expert discourse. In a highly technological modern society with complex policy-making patterns it is, however, important to ask in what social arena and how the values and goals can be mutually understood by experts and laypeople. How is the civil society mobilised to protect nature or to secure a sustainable living environment?

According to our view the responsibility of sustainable environmental policy cannot simply be transferred to experts or environmental movements from the community level. Civil society should create particular rational models of self-regulation to guarantee a social arena for risk perception, for dealing with values and preferences, and for democratic choice of policy action. This would be an arena where experts and laypeople could meet on an euqal basis. Personal impact assessment and ecosocial profiles built and represented in everyday life would, then, serve as social resources to increase the self-regulative capacities of the society.

The personal mode of action commonly follows the model of teleological reasoning which may be composed as follows: (1) the actors recognises a factual situation that has to do with a highly appreciated object. (2) Having realised the factual situation, the actor chooses another situation in relation to the appreciated object. The more preferred situation is set as the ultimate goal of action. (3) The actors then pursues this goal by specific means that are considered to be essential for achieving the goal. For example, in the case of the accelerated greenhouse effect, the members of the ecosocial moral community can choose appropriate means in their daily lives to prevent the increase of carbon dioxide emissions and to join in this way a more global strategy of rational action to mitigate environmental risk. This kind of reasoning is structured in the way that usually is recognised as an ideal model of expert discourse, no matter how much the actual mitigation may differ from the ideal model. As an ideal model, it may, however, serve as an analysis of the gap between the expert and the lay perspective on the issue of sustainability of the environmental and environmental risk.

As said before, the formulation of the ecosocial community in civil society is usually actualised through a policy of small actions: there is rarely involved any direct structural changes at the level of state policies due to these actions. However, communication between experts, professionals, and laypeople can be considered a desired outcome of the development of ecosocial communities. The personal impact assessment and ecosocial profiles of individual families provide a channel of information which is likely to offer new resources for environmental planning and administration, if this agency of the civil society can be channeled to the societal management processes, such as environmental impact assessments.

There can be drawn a crude distinction between (1) a strategy of rational political influences and (2) a mythical survival strategy. It is commonly assumed that experts and professionalists are inclined to follow a strategy of rational environmental influence and that laypeople are inclined to resort to a mythical environmental survival strategy. The baseline assumption is that there is an important difference in the strategic approach or in the form of rationality between experts and laypeople. It is likely that a pure division of social values and action orientation, such as presented by the model of "theoretical teleologies" (see Figure 12.1), has never been found historically. A more reasonable approach is to argue that there is a large scale of variation around the ideal types. Both experts and lay people can and do combine elements of both rational and mythical analysis in their strategic

approaches. Yet it is more probable that we find the ideal types when analysing the strategic behaviour in different social fields, as dominating strategies on each side. To facilitate communication between two relatively distant partners, it is of general interest to develop methods of auditing and participation by civil society such as the environmental impact assessment. There is, however, no technological solution to the problem of social distance between experts and laypeople. This is why environmental impact assessment needs to be completed and reformed by social impact assessment.

Environmental impact assessment was institutionalised rapidly in the 1980s as an internationally recognised strategy for environmental policy. Since the 1980s, EIA procedures have developed into multiple strategies and vary by country. As a process, EIA may, however,be defined by its phases. According to Glasson, the main steps of the EIA process can be defined as in Table 12.1.

**Table 12.1**

| *Step* | *Tasks* |
|---|---|
| 1. | Project screening (is an EIA needed?)<br>Scoping (which impacts and issues to consider?)<br>Description of the project/development<br>Description of the environmental baseline<br>Identification of key impacts |
| II | Prediction of impacts<br>Evaluation and assessment of significant of impacts<br>Identification of mitigating measures |
| III. | Presentation of findings in an Environmental Impact Statement (EIS) |
| IV. | Review of the EIS |
| V. | Decision making |
| VI. | Postdecision monitoring<br>Auditing of predictions and of mitigation measures. |

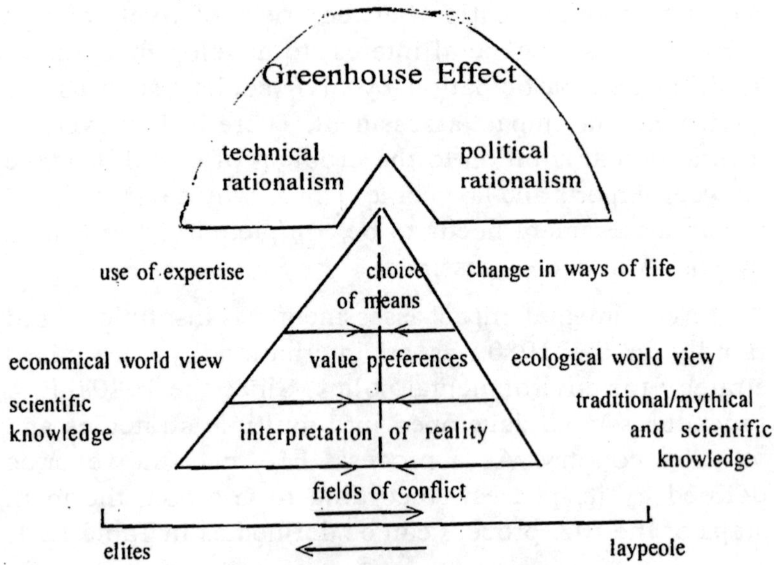


**Figure 12.1: Theoretical Teleologies**

Public consultation is a formal part of the EIA, which may be attached to several phases of the process. More important here is that this involves the official recognition of the "parties interested" in each EIA process *separately*. From the point of view of sustainable development, however, a good environmental standard may be defined to be a *general interest*. Consequently, it is important to discuss in what framework this general interest can be presented.

From a sociological point of view, EIA has recently been introduced with reference to reflective modernisation, society risk, and communicative planning. It is assumed to develop into a tool of environmental management in order to promote ecological modernisation of a society. A critical view of EIA is concerned with the introduction of an effective shift from scientific-data-based impact

assessment to a value-based assessment of environmental change. It also calls strongly for the mobilisation of the social action resources of the civil society, and, furthermore, it emphasises developing linkages between lay people's perceptions and expert knowledge.

While *environmental* impact assessment has focused mainly on the physical and biological properties of natural environments, Social Impact Assessment (SIA) concentrates on the distinctively human side of particular environments. The very heart of the SIA is to examine the social impact of formulating policies, instituting programmes, and building projects. As does the environmental impact assessment, SIA aims to predict and evaluate the impact and outcomes of planned actions. This is, however, approached from the individual, family, or community point of view. SIA methods have been developed since the 1970s in parallel with EIA methods, but often with a distance and distinctiveness from the development of EIA expertise. Today the problem of the (re) integration of EIA and SIA appears to be one of the basic questions not only from a pragmatic policy point of view, but also when considering ethical aspects of the more indirect human impacts related to development acts.

According to its basic idea, SIA uses local citizens' environmental awareness as an input to the societal process of environmental impact assessment. The SIA process should bring forth the shared or controversial values of the community and create a value-based process of mitigation in the interest of the local community. There are several alternatives of practical strategies for the SIA. Auditing may be part of the institutionalised process of EIA, citizens may be involved in the debate concerning EIA in mass media, or a local future workshop an be created. We suggest one more action idea that could be practical as a more permanent participative appraisal for ecosocial communities.

There have been, e.g., in Finland, some recent tentative experiences of integrating social work in local communities with the zoning processes of the cities in a way which we find to be a very interesting example of a new community-level policy of sustainable development. The ecosocial approach to social work pursues in this case an empowerment strategy, trying to match city planning to the ecosocial profiles and projection of the local population. The conventional experts of the EIA process are involved, of course, but there is an additional contribution by the local social workers to mediate between experts and lay people. In this case, social workers are developing capacities as negotiating professionals in ecosocial mitigation. By outlining community values to the experts, they help the experts to determine the relevant criteria of a good living environment. On the other hand, the experts' knowledge and technical resources can be translated to the members of the community. The application of this two-way process of advocacy and empowerment calls of an open professionalism where traditions of social closure may be avoided among social workers. The capacities of the reflexive and open professionalism are tested particularly in controversial cases such as in construction of new roads or supermarkets at the next neighbourhood of the ecosocial community.

We mention the example of the Finnish social workers as a particular application of SIA mainly because their experiences have further clarified some basic elements of the lay perspective on the immediate living environment. The social workers' activity as SIA agents also increased our understanding of what kind of social mechanisms are applicable to channel laypeople's ethical purposes to the EIA process in order to promote sustainable development.

The most crucial qualities or basic values of the urban living environment were brought to the fore through the process of reciprocal participation between citizens and experts. The SIA interpretation by the social workers of the vital value basis refers to the preference for three qualities of an urban environment:(1) good quality or state of the physical and biological environment, (2) social and ecological diversity of the environment, and (3) facility of quotidian life. The ecosocial community is, of course, expected to appreciate highly qualities (1) (2), whereas the third element is not as easily assumed as part of the quality of a sustainable environment. From the community's point of view, this aspect may, however, be very decisive in everyday life when people choose the particular acts of their ecosocial policy of small actions. Thus, the final outcome of the SIA process should be balanced as an assessment of the man-nature relationship and should include even the feedback the man-to-man relationship has on the former relation.

Environmental problems are inherently social problems. They have been generated mainly by the industrial societies and by consumers' way of life. Hence, it is consistent that particularly these societies create strategies of social action aiming at mitigation of these problems. Environmental impact assessment has been developed into a legitimate tool for experts to define the impacts of environmental change. In a modern society, ecological development is, however, a major social issue in the everyday life of the common people. Daily choices by individuals and communities do have global consequences. This is why environmental impact assessment ought to be reformed and complemented by social impact assessment.

Social impact assessment involves communication not only from expert to expert but also between experts and lay people. It is ultimately the mission of the civil society to create a culture of self-control in relation to the living environment. In highly individualised societies—and in the world of globalised risk for the environment—the social resources of sustainability can be restored to a great extent to the moral obligations of the solidarity the citizens feel for the "distant other," whether these others are represented by faraway people or by people of future generations. The new ecosocial morality of small steps of environmental protection in everyday life is not to be minimised but to be recognised as an element of a new culture promoting ecosocial morality in civil society. The more communities are empowered by the new ecosocial morality, the stronger the capacity they have in generating political impulses to the process of environmental impact assessment in order to support future prospects of sustainable development.

## REFERENCES

Bourdieu, P. (1987), *Choses dites,* Paris: Les Editions, de Minuit.

Douglas, M., & Widasky, A. (1983), *Risk and Culture, an Essay on the Selection of Technological Dangers,* Berkeley, University of California Press.

Finsterbusch, K., Llewellyn, L.G., & Wolf, C.P. (eds.), (1993), *Social Impact Assessment,* Beverly Hills, CA: Sage.

Hofbeck, J. (1991), La Deep Ecology, Essai d'evaluation ethique. In J.A. Prades, J.G. Vaillancourt, et R. Tessier (Eds.), *Environmental et development: Questions ethiques et problems socio-politiques,* Quebec: Editions Fides.

Morin, E., & Kern, A. (1993), *Terre-Parrie,* Paris: Seuil

Notton, B. (1987), *Why preserve Natural Variety?* Princeton: Princeton Universivy Press.

Ollikainen, M. (1992), Kestävä kehitys: Talouden ekologisointi eettisten ja taloustieteellisten periaatteiden nojalla (Sustainable development: Ecological perspective on economy with reference to principles of ethics and economy). Yhteiskunta ja ymparisto, Ympäristopolitikan arvo-ongelmia M. Järvelä

ja Y. Uurtimo (Eds.), Julkaisuja Yhteiskuntatieteiden tutkimuslaitos, Tampereen yliopisto 2/1992, 101-123.

Redclift, M. (1987), *Sustainable Development.* London: Methuen.

Sairinen, R. (1993), Environmental Impact Assessment in Finland. In A. Haila (Ed.), *Cities for Tomorrow—Directions for Change.* Espoo: Helsinki University of Technology, Centre for Urban and Regional Studies.

Tessier, R. (1991), L'ethique du developpement durable: Quels fondements? Une comparaison avec l'ascetisme seculier chez Weber. In J.A. Prades, J.G. Vaillancourt, et R. Tessier (Eds.), *Environment et Developpement. Questions ethiques et problems socio-politiques,* Quebec: Editions Fides.

Wolf, C.P. (1983), Social Impact Assessment: A methodological overview, In K. Finsterbusch, L.G. Llewellyn, and C.P. Wolf (Eds.), *Social Impact Assement,* Beverly Hills, CA: Sage.

# BIBLIOGRAPHY


Arnold, Steven H., 1989, "Sustainable Development: A Solution to the Development Puzzle," *Development, Journal of SID* 2/3:21-24.

Bell, Daniel, 1976, *The Cultural Contradictions of Capitalism,* New York: Basic Books.

——— ,1977, "Are There 'Social Limits' to Growth?" In Kenneth D.Wilson, ed., *Prospects for Growth: Changing Expectations of the Future,* New York: Praeger.

Bharadwaj, I., 1984, "The Contribution of Gandhian thinking to the Understanding of the World Crisis," *Revue internationale de sociologie* 20:65—99.

Borgatta, Edgar F., 1989, "Towards a Proactive Sociology," Paper presented at the 29th International Congress of the International Institute of Sociology, Rome.

———, and Karen S. Cook, 1988, "Sociology and Its Future." In Edgar F. Borgatta and Karen S. Cook, eds., *The Future of Sociology.* Newbury Park, Calif.: Sage.

———, and Laurie Russell Hatch, 1988, "Social Stratification," In Edgar F. Borgatta and Karen S. Cook, eds., *The Future of Sociology,* Newbury Park, Calif. Sage.

Brown, Lester R., Christopher Flavin, and Sandra Postel, 1990, "Picturing a Sustainable Society," In Lester

R. Brown and Associates, *State of World 1990: A Worldwatch Institute Report on Progress Toward a Sustainable Society.* New York: W. Norton.

Brundtland, Gro Harlem, 1989, "Sustainable Development: An Overview. *Development, Journal of SID* 2/ 3:13-14.

Buttel, Frederick H. 1987 "New Directions in Environmental Sociology," *Annual Review of Sociology,* 13:465-488.

Catton, William R., Jr. 1978, "Carrying Capacity, Overshoot, and the Quality of Life." In J. Milton Yinger and Stephen J. Cutler, eds., *Major Social Issues: A Multidisciplinary View,* New York: Free Press.

———, 1980, *Overshoot,* Urbana: University of Illinois Press.

———, Jr., and Riley, E. Dunlap, 1978 "Environmental Sociology: A New Paradigm." *American Sociologist,* 13:41-49.

Commoner, Barry, 1974, "Interview on Growth," In Willem L. Oltmans, ed., *On Growth: The Crisis of Exploding Population and Resource Depletion,* New York: Capricorn Books.

Coomarasway, A.K., 1946, *The Religious Basis of the Forms of Indian Society,* New York: Orientalia.

de la Court, Thijs 1990, *Beyond Brundtland: Green Developments in the 1990s.* New York: New Horizon Press.

Duncan, Otis Dudley, 1964, "Social Organisation and the Ecosystem." In Robert E. L., Faris, ed., *Handbook of Modern Sociology.* Chicago: Rand McNally.

Dunpal, Riley E., 1979, "Environmental Sociology." *Annual Review of Sociology,* 5:243-273.

———, 1986, "Two Decades of Public Concern for

Environmental Quality: Up, Down and Up Again." Paper presented at the Annual Meetings of the American Sociological Association, New York.

Ehrlich, Paul R., 1968, *The Population Bomb*, New York: Ballantine Books.

Fench, Hilary F., 1990, *Green Revolutions Environmental Reconstruction in Eastern Europe and the Soviet Union,* Worldwatch Paper 99:1-62. Washington, D.C.: Worldwatch Institute.

Freudenberg, William R., and Robert Gramling, 1989, "The Emergence of Environmental Sociology: Contributions of Riley E. Dunlap and William R. Catton, *Jr." Sociological Inquiry,* 59: 439—452.

Gottdiener, M., and Joe R. Feagin, 1988, "The Paradigm Shift in Urban Sociology," *Urban Affais Qualterly* 24:163—187.

Hardin, Garrett 1968 "the Tragedy of the Commons." *science* 162: 1243-1248.

Hawley, Amos, H.,1979, "Cumulative Change in Theory and in History." In Amos H. Hawley, ed., *Societal Growth: Processes and Implications,* New York: Free Press.

———,1984, "Sociological Human Ecology: Past, Present, and Future," In Michael Micklin and Harvey, M. Choldin, eds., *Sociological Human Ecology: Contemporary Issues and Applications.* Boulder, Colo: Westview.

———, 1986, *Human Ecology: A Theoretical Essay.* Chicago: University of Chicago Press.

Hilidyard, Nicholas B., 1989, "Adios Amazonia? A Report from Altimira Gathering." *The Ecologist,* 19:53-67.

Holden, Constance, 1989, "Environment, Culture, and Change in the Arctic," *Science,* 243:883.

Humphrey, Craig R., and Frederik H. Buttel 1982 *Environment, Energy, and Society,* Belmont, Calif: Wadsworth.

Kahn, Herman, 1974, "Interview on Growth." In Willem L. Oltmans, ed., *On Growth: The Crisis of Exploding Population and Resource Depletion.* New York: Capricorn Books:

Leonard, Pamela, 1990, *Effective Gobal Environmental Protection: World Federablist Proposals to Strengthen the Role of the United Nations,* Washington, D.C.: World Federalist Association. (In Collaboration with Walter Hoffman.)

Meadows, Donella, Dennis Meadows, Jorgen Randers, and William Behrens, 1972, *The Limits to Growth: A Report for the Club of Rome's Project on the Predicament of Mankind,* New York: Universe Books.

Medvedev, Zhores A., 1990, "The Environmental Destruction of the Soviet Union," *The Ecologist* 20: 27-29.

Mellos, Koula, 1988, *Perspectives on Ecology: A Critical Essay,* New York: St. Martin's.

Micklin, Michael, and Harvey, M. Choldin (eds.), 1984, *Sociological Human Ecology: Contemporary Issues and Applications,* Boulder, Colo.: Westview.

Miller, G.Tyler, Jr. 1972, *Replenish the Earth: A Primer in Human Ecology. Belmont* Calif.: Wadsworth.

Morrison, Denton E., 1976 "Growth, Environment, Equity and Scarcity," *Social Science Quarterly,* 57: 292-306.

Morrison, Denton E.,1978, "Equity Impacts of Some Major Energy Alternatives." In Seymour Warkov, ed., *Energy Policy in the United States: Social and Behavioural Dimensions.* New York: Praeger.

———— ,1980, "The Soft, Cutting Edge of Environmentalism: Why and How the Appropriate Technology Notion Is Changing the Movement." *Natural Resources Journal* 20:275-298.

————, and Riley E. Dunlap, 1986, "Environmentalism and Elitism: A Conceptual and Empirical Analysis." *Environmental Management* 10:581-589.

Ophuls, Willam, 1977, "Buddhist Politics: In Politics as in Economics, Small Is Beautiful," *The Ecologist,* 7:82-87.

Park, Robert E., 1936, "Human Ecology," *American Journal of Sociology,* 42:1-15.

Polanyi, Karl, 1974, "Our Obsolete Market Mentality," *The Ecologist,* 4:213-220. Reprint of 1947 article.

Rappaport, Roy A., 1976, "Forests and Man," *The Ecologist,* 6:240-246.

————, 1978, "Biology, Meaning, and the Quality of Life." In J. Milton Yinger and Stephen J. Cutler, eds., *Major Social Issues: A Multidisciplinary View,* New York: Free Press.

Redclift, Michael, 1987, *Sustainable Development: Exploring the Contradiction,* New York: Methuen.

Reichel-Dolmatoff, G., 1977, "Cosmology as Ecological Analysis: A View from the Rain Forest," *The Ecologist,* 7:4-11.

Revelle, Roger, 1974, "Interview on Growth," In Willem L. Oltmans, ed., *On Growth: The Crisis of Exploding Population and Resource Depletion,* New York: Capricorn Books.

Sachs, Wolfgang, 1990, "On the Archeology of the Development Idea." *The Ecologist,* 20:42-43.

Saran, A.K., 1977, *Gandhi and the Concept of Politics: Toward a Normal Civilisation,* New Delhi: Nehru Memorial Museum and Library.

———, 1978, "The Traditional Vision of Man." UNESCO Seminar, Hyderabad, India, Mimeo.

Schnaiberg, Allan 1980 *The Environment: From Surplus to Scarcity,* New York: Oxford Unviersity Press.

——— 1983 "Redistributive Goals Versus Distributive Politics: Social Equity Limits in Environmental and Appropriate Technology Movements," *Sociological Inquiry* 53-200-219.

Sen, Amartya K., 1981, *Poverty and Famines: An Essay on Entlement and Deprivation,* Oxford: Clarendon Press.

Singh, Digvijay, 1985, *The Eco-Vote: People's' Representative and Global Environment.* New Delhi: Presntice-Hall of India.

Sinha, Radha, 1976, *Food and Poverty: The Political Economy of Confrontation,* London: Croom Helm Ltd.

Smith, Michael P., 1979, *The City and Social Theory,* New York: St. Martin's.

Stonich, Susan C., 1989, "The Dynamics of Social Processes and Environmental Destruction: A Central American Case Study." *Population and Development Review* 15:269-296.

Tokar, Brian, 1988, "Social Ecology, Deep Ecology and the Future of Green Political Thought," *The Ecologist,* 18:132-141.

Vohra, B.B. 1985, "Why India's Forests Have Been Cut Down.," *The Ecologist,* 15:50-51.

World Commission on Environment and Development, 1987, *Our Common Future,* New York: Oxford University Press (The Brundtland Report).

Young, Genrald (ed.), 1983, *Origins of Human Ecology. Stroudsburg,* Pa.: Hutchinson Ross.

# INDEX